Alexander of Aphrodisias
On Aristotle's *Meteorology 4*

Alexander of Aphrodisias

Aphrodisias

On Aristotle's
Meteorology 4

Translated by Eric Lewis

Cornell University Press

Ithaca, New York

Introduction and translation © 1996 by Eric Lewis
Appendix © 1996 by Richard Sorabji

First published 1996 by Cornell University Press.

Library of Congress Cataloging-in-Publication Data

Alexander of Aphrodisias.
   On Aristotle's Meteorology 4 / Alexander of Aphrodisias ;
translated by Eric Lewis.
      p.   cm.
   Includes bibliographical references and index.
   ISBN 0-8014-3225-1 (alk. paper)
   1. Meteorology–Early works to 1800. 2. Aristotle.
Meteorologica. 3. Alexander, of Aphrodisias–Contributions in
meteorology.   I. Title.
QC859.A44 1995
551.5–dc20   95-37120

Acknowledgements

The present translations have been made possible by generous and
imaginative funding from the following sources: the National Endow-
ment for the Humanities, Division of Research Programs, an
independent federal agency of the USA; the Leverhulme Trust; the
British Academy; the Leventis Foundation; the Humanities Research
Board of the British Academy; HEFCE; the Wolfson Foundation; the
Jowett Copyright Trustees; the Royal Society (UK); Centro Internaz-
ionale A. Beltrame di Storia dello Spazio e del Tempo (Padua); Mario
Mignucci; Liverpool University. The editor wishes to thank D. Russell,
R. Sharples, D. Furley, M.L. Gill, J. Lennox, D. Sedley, and P. Lautner
for their comments on the translation; I. Crystal for preparing the
Greek-English index; and R. Wisnovsky for his help in preparing the
volume for press. He also thanks Dr Coutant for his permission to put
his earlier translation to use.

Printed in Great Britain

Contents

Introduction

> That the *Meteorologica* is a little-read work is no doubt
> due to the intrinsic lack of interest of its contents. Aristotle
> is so far wrong in nearly all his conclusions that they can,
> it may with justice be said, have little more than a passing
> antiquarian interest ... The main interest of the work is
> to be found not so much in any particular conclusions
> which Aristotle reaches, as in the fact that all his conclu-
> sions are so far wrong and in his lack of a method which
> could lead him to right ones. In this he is typical of Greek
> science.[1]

This passage comes from the introduction to H.D.P. Lee's 1952
translation of the *Meteorologica* in the Loeb Classical Library,
and was allowed to stand in the second edition of 1962. That
such an eminent scholar of this text could be so disparaging
about it might give us pause before delving into it, let alone
into an ancient commentary on it. Indeed, one might be doubly
wary, since the authenticity of Book 4, the work we are concerned with here, has been called into question. In the following
essays I hope to convince the reader that *Meteor.* 4 is both
genuine and of great, indeed central, interest to one concerned
with Aristotelian natural philosophy and metaphysics. Establishing the importance of *Meteor.* 4 itself will at the same time
establish the interest of ancient commentaries on it. Further
sections will concentrate on what is of most interest within
Alexander's commentary.

The overt positivism of Lee's comments points to his particular prejudice, a disproportionate concern with the correctness
of the scientific doctrines found in the *Meteorologica*. Even

---

1. H.D.P. Lee (trans.), Aristotle, *Meteorologica*, Loeb Classical Library, Cambridge,
Mass. 1952 (second edition, 1962), xxv-xxvi.

without entering into debates concerning the proper attitude
to take when critically judging non-contemporary science, it is
clear that the doctrines themselves, and in particular whether
they are 'true' (itself a slippery issue), do not come close to
exhausting the interest of Aristotelian scientific works. The
recent explosion of interest in Aristotle's biological works,
equally full of 'false conclusions', and the great advances in our
understanding of central concepts in Aristotle's metaphysics
and natural philosophy that such studies have engendered,
should lay to rest reasoning such as Lee's.[2] Were we to discount
the biological corpus for its antiquated biology, our understanding of the rest of the corpus would be substantially
impoverished.

I hope to show that *Meteor.* 4 has the same importance as
works from the biological corpus are now recognised to have.
Indeed, I shall attempt to establish a rather close link between
Meteor. 4 and the biological corpus. But more importantly, I
shall argue that in *Meteor.* 4 Aristotle modifies, in radical ways,
doctrines central to his physics. In particular, I will show how
Aristotle's views concerning form, matter and the elements
undergo a radical transformation in *Meteor.* 4, and that this
work should therefore be considered a fundamental text in the
Aristotelian corpus.

Such claims for the importance of *Meteor.* 4, and in particular
for its revisionary nature, bring the issue of authenticity to the
fore. Perhaps *Meteor.* 4 is so revisionary precisely because it is
not by Aristotle. Therefore I will start by discussing the question of authenticity. In a sense all the sections that follow are
an extended argument for the work's authenticity, since they
show just how interrelated it is with many other Aristotelian
texts of unquestioned pedigree. I will then discuss the doctrines
most central to the work, with the aim of establishing precisely
its innovative nature. Often my notes to the translation will
refer to this introduction.

2. Contrast Lee's view with that of G.E.L. Lloyd, who recognises the importance of
Meteor. 4, telling us that in it we 'have, for the first time in Greek science, an attempt
to discuss a quite wide range of physical changes and phenomena' (*Magic, Myth and
Experience*, 209).

1. The authenticity of *Meteor*. 4, and the question of pores

The debate concerning the authenticity of *Meteor*. 4 (unquestioned until the turn of this century), has as its locus classicus I. Hammer-Jensen's 1915 paper 'Das sogenannte IV. Buch der Meteorologie',[3] where it is argued that *Meteor*. 4 is spurious, in fact having Strato as its author. Ross and Jaeger agree with her as to the spuriousness of the work, while refusing to attribute it to a particular author.[4] Its authenticity was then defended by H.D.P. Lee and I. Düring,[5] but in 1961 it was again questioned by H.B. Gottschalk, who claimed that it is by Theophrastus, or at least extracted from a work of his.[6] The debate seems to come to a close in 1983, with D. Furley's rebuttal of Gottschalk,[7] arguing that Aristotle is, after all, the author of *Meteor*. 4.

One essential person has been left out of this debate, and that is Coutant. Gottschalk mentions him as someone who has defended the authenticity of the text, but he acknowledges in a footnote that he has 'not been able to see this book'.[8] Furley mentions Coutant in one footnote, yet chose not to mention any of his arguments in favour of authenticity, although they are quite congenial to those that Furley himself produces.[9] In what follows I will sketch the main strands of the debate. I am in agreement with many of the arguments put forth by Coutant and Furley, and in the rest of the Introduction I will build upon some suggestions made by both.

Jensen's claim that *Meteor*. 4 is by Strato can be rebutted on chronological grounds. As Coutant notes, Aristotle at least saw

---

3. In *Hermes* 50, 113-36.
4. Jaeger, *Aristoteles*, 412f.; Ross, *Aristotle*, 11.
5. Lee, op. cit., xiii-xxi; Düring, *Aristotle's Chemical Treatise Meteorologica Bk IV*, Goteburg 1944.
6. H.B. Gottschalk, 'The Authorship of Meteorologica, Bk IV', *Classical Quarterly* N.S. xi, 1961, 67-78. Lee, in the second edition of his Loeb of the *Meteor*., is unconvinced by Gottschalk's arguments.
7. 'The Mechanics of *Meteorologica* IV', 148 of reprint in *Cosmic Problems*.
8. Gottschalk, op. cit., 67, n. 4.
9. For instance, Furley summarises approvingly Gottschalk's arguments against Jensen's thesis that Strato is the author of *Meteor*. 4, while the very same arguments are found much earlier in Coutant. See Furley, op. cit., 132, and Coutant, op. cit., 9.

Meteor. 4, and referred to it,[10] and so it must have been written before his death. 'In that case, we have Strato, who lived fifty years after Aristotle, writing, probably in his early twenties, a work which Aristotle referred to several times for fuller treatment of a topic mentioned fleetingly, so impressed was he by its high quality. This is improbable on the face of it.'[11] Furley concurs, forcefully stating that '... book IV is referred to in genuine Aristotelian treatises that were in all probability written before Strato was old enough to read, let alone write book IV ...'.[12] It is with the doctrines found in book 4 that the real case for or against authenticity rests.

The central argument against authenticity, propounded by both Jensen and Gottschalk, concerns the theory of pores found in 4.8-9. As Gottschalk puts it, 'Aristotle insisted that all material substances are continuous and homogeneous ... but in [*Meteor.* 4] we find a theory that bodies are riddled with tiny channels or "pores" through which one substance can penetrate and act upon another'.[13] As for the possibility that Aristotle might have imagined the existence of pores to facilitate certain chemical interactions, Gottschalk tells us that 'Aristotle did consider this idea, and his words leave no room for doubt as to his opinion. He called it false, superfluous, and ridiculous. Nor is it likely that he should have changed his mind.'[14]

Before turning to the theory of pores as found in *Meteor.* 4, and the arguments against pores found in *Phys.* 4.9 and *GC* 1.8, let us examine in general Gottschalk's claims. How likely is it that Aristotle thought that no matter is penetrated by holes? Did Aristotle rule out the existence of Swiss cheese, or that common Aegean animal, the sponge, on *a priori* grounds? Of course not, and in fact Aristotle's extended discussion of the sponge (*HA* 5.16) repeatedly claims that they are 'loose in texture' (*manos*), and, conclusively, that they contain pores, even vacant pores (*poroi kenoi*). Gottschalk, as Furley realises, is misconstruing the sense in which Aristotle takes material substances to be both continuous and homogeneous. Matter is

---

10. Coutant, 9.
11. Coutant, 9.
12. Furley, 132.
13. Gottschalk, 68.
14. Gottschalk, 71.

continuous for Aristotle in the sense that it is not atomic, it is not composed of small, discrete, indivisible building blocks, themselves either interlocking mechanistically, or separated by void.[15] Matter is homogeneous for Aristotle in the sense that things which in fact are homogeneous have parts like the whole in terms of composition. Yet, as Furley states, Aristotle neither thinks that all matter is homogeneous, nor does he take homogeneity to be always strictly applied. 'Aristotle's discussion of blood in *PA* 2.4 shows that in his view a substance may be continuous, and homeomerous,[16] and yet contain some parts that are solid (*hai kaloumenai ines*) and some that are liquid.'[17]

There is another, more obvious, sense in which Aristotle denies homogeneity in the manner in which Gottschalk would apply the term. For Aristotle, as for any reasonable theorist, substances are only homogeneous, if they are homogeneous at all, *where they are*. It is no use considering a bit of Swiss cheese and a hole in it, and denying that the Swiss cheese is homogeneous on the grounds of the hole, for the Swiss cheese contains the hole, but not as a part. The Swiss cheese is topologically complex, which is perfectly consistent with being homogeneous, since homogeneity for Aristotle is a compositional, not a topological, notion. Whether some bit of matter is homogeneous for Aristotle is a question answered by looking at what it is made of, and precisely how these ingredients are incorporated into the whole. The shape of the whole does not come into play. One might as well deny the compositional homogeneity of the cheese by considering a bit of the cheese and the china plate it rests upon. It is clear that such criticisms will not do.

Aristotle's criticisms of pores, in both *Phys.* and *GC*, are meant to deny two general claims. First, that pores are necessary to account for the divisibility of bodies, and secondly, that pores, and the purely mechanistic movement of atoms through them, are sufficient to explain all qualitative alterations that a body may undergo. This leaves open the dual possibilities that pores of some sort are involved in an explanation of why some physical body is physically divided where it is (yet are not

15. The sense in which the atoms were thought to be indivisible, both by Aristotle and the atomists themselves, is hotly debated.
16. A term meaning 'having parts all like'.
17. Furley, 143.

necessary in order for such a body to be divisible), and that passage through pores might be necessary in order for some qualitative changes to take place (yet not sufficient for bringing it about). *Meteor.* 4 affirms both of these possibilities, which are consistent with the objections to pores found in both *GC* and *Phys.* Let us look at these objections more closely.

As Coutant notes with *GC* 1.8 in mind (and Furley notes with respect to *Phys.* 4.9), the type of pores that Aristotle is at pains to discredit are 'pores as absolute voids (the atomistic pores), which, according to the Atomists, are the avenues of sensation'.[18] Furley is surely right in claiming that 'so long as we accept that *Meteor.* 4's pore-theory does not include the existence of void, there is nothing that is inconsistent with *Phys.* IV on this score'.[19]

It is clear that the attacks on pores found in *Phys.* all indirectly ride piggy-back on the attacks on void. The more direct attack on pores in *GC* 1.8 bears somewhat closer scrutiny. It is the closing passage of 1.8 which is, according to Gottschalk, inconsistent with *Meteor.* 4, and so I will quote it in full:

> From what has been said it is apparent that the postulation of pores, as some entertain them, is either false (*pseudos*) or superfluous (*mataion*). And since bodies are divisible throughout, to assume pores exist is ridiculous (*geloion*), for that which is divisible [everywhere] can be divided [anywhere] (1.8, 326b24-8).

Aristotle claims that the postulation of pores, *as some entertain them* is either false, superfluous, or ridiculous. We are meant to know why from what precedes, yet it is clear that we are not faced with a general renunciation of all possible pore theories. Contrary to Gottschalk's opinion, the aforementioned three objections to pore theories are not generic. Aristotle is not slandering pore theories by claiming that they all are false, superfluous and ridiculous, but certain pore theories, formulated in precise ways, are either false, superfluous, or ridicu-

18. Coutant, 14.
19. Furley, 141.

lous. How so? From the body of *GC* 1.8 it is clear that pore
theories that assume that pores are void spaces are false, those
that reduce all qualitative change to contact via pores are
superfluous, and, as we are told in the above quote, those that
assume that pores are needed to ground the divisibility of
bodies are ridiculous.

Can the 'pore theory' of *Meteor*. 4 meet these three charges?
The first charge is the easiest to deal with. As mentioned above,
it is clear that the pores of *Meteor*. 4 are not void spaces. As for
the third charge, a full discussion of the 'type' of indivisibility
the atoms of Democritus and Leucippus enjoy, and of Aristotle's
interpretation of atomism, is needed. Yet it is clear from the
passage quoted that Aristotle thinks that pores are ridiculous
if invoked to explain why bodies are divisible, since Aristotle
thinks that in some sense or other, all bodies are divisible
everywhere (and so they would have to be everywhere void if
void was necessary for divisibility). When he claims that bodies
are divisible everywhere, he denies that this feature of bodies,
or magnitudes in general, is due to the existence of pores (how
could it be if this feature of bodies is said to be due merely to
the fact that bodies are magnitudes?). Yet this denial is perfectly consistent with claiming that some particular physical
divisions of some physical bodies are due to their physical
structure, and in particular due to the structure of pores in
them. This is all that *Meteor*. 4.9 claims in its account of things
that can be broken, as opposed to those that can be shattered.
All Aristotle needs still to believe while writing *Meteor*. 4 is that
the manner in which, say, a certain piece of wood is in fact
physically split, does not entail that it is divisible only in this
way, and there is no reason to think that the Aristotle of *Meteor*.
4 thought otherwise.

The second charge that the postulation of pores is superfluous, requires closer examination. Aristotle is objecting to an
atomistic theory of interaction, which reduces all qualitative
change on the part of bodies to rearrangement of the atoms of
which they are composed. The different ways in which atoms
can be arranged, or can be in contact, accounts for the different
properties bodies can have, and the different processes they can
undergo. An essential part of this atomist theory of qualitative
change is that the atoms can differ one from another only with

respect to size, shape and motion. Atoms do not differ one from another with respect to any 'intrinsic' properties, nor do any of their properties change when they interact, apart from their motion. Consequently, Aristotle claims that if a qualitative change takes place in a macroscopic body merely by means of the atoms it is composed of coming to be in contact with other atoms in a new way, then it should make no difference whether this contact takes place solely on the outer surface of the body, or within it via pores running through it.

This atomist model of qualitative interaction is tacitly being held in opposition to Aristotle's view as developed in *GC*, which has it that the amount of surface area that two bodies have in contact affects whether a real qualitative change in the objects may take place. If, according to the atomists, there are no 'real' qualitative changes, but only rearrangements of atoms, then, so Aristotle thinks, external contact should suffice in all cases, making pores otiose.[20] 'If [an agent] does not act according to contact, neither will it act by passing through pores.'[21] For Aristotle contact is one of the necessary preconditions for two bodies to interact qualitatively, but qualitative change itself is not reduced to changing configurations of bodies in contact. Qualitative change merely commences when two bodies suitable for interacting come to be in contact in the appropriate way.

Since Aristotle's own positive theory of qualitative interaction involves interacting objects' being in contact, and so moving in order to come into contact, the theory has a mechanistic element. So it should come as no surprise, and raise no suspicions of inauthenticity, to discover that many of *Meteor*. 4's accounts of assorted affections of bodies, by far the most detailed in the whole Aristotelian corpus, involve descriptions of the movements of things interacting.[22]

20. This criticism of Aristotle's, often thought to be decisive, is in fact flawed. If atomism has it that the qualities of macroscopic bodies 'emerge' out of the micro-atomic structure of such bodies, then there is every reason to think that many emergent properties will supervene on atomic configurations which involve the intertwining of atoms not just at the surfaces of those macroscopic bodies that they are the atoms of, and so mere external contact would not be sufficient to account for all apparent qualitative changes. In other words, even if all qualitative interaction is reduced to the rearrangement of atoms, the appropriate rearrangements may require the interweaving of the atoms of the objects which are involved in a qualitative interaction.

21. *GC* 1.8, 326b21-2.

22. These may be the bodies themselves interacting, or, as is more often the case, the elements, or contraries, these bodies are composed out of.

In so far as qualitative interactions are somewhat, or sometimes, dependent on the amount of surface the interacting objects have in contact, Aristotle's positive theory invites the introduction of pores in bodies (since pores allow more surface of the interacting bodies to be in contact). And we do find pores introduced to explain melting, softening, splitting, shattering, compressing, combustion, and fission. Again, this should raise no eyebrows.

To summarise, what Aristotle is objecting to is a theory of qualitative change which reduces all such changes to topological changes, and reduces all qualities to topological features. In particular he is objecting to atomism, which also invokes void spaces, sometimes pores, in such reductions. The account of qualitative change found in *Meteor*. 4 is not reductionist in the way in which it is in atomism, nor does it involve void pores, and so it is, on these accounts at least, perfectly consistent with *GC* and *Phys*.

2. The place of *Meteor*. 4 in the corpus

Having defended the authenticity of *Meteor*. 4, I will now turn to the question of its place in the corpus. That it does not belong to the rest of the *Meteor*. is almost universally agreed. Alexander places the work with *GC*, although he does not think that it is a part of that work.[23] Olympiodorus places it between *Cael.* and *GC*,[24] while Patrizi opts for it preceding *PA*.[25] Lee, following Coutant, considers it to be 'a separate treatise'.[26] Furley, while agreeing with most that it does not belong with *Meteor*. 1-3, claims that '... there is a sense in which it is even in the right place ... the introduction to *Meteorologica* announced that biology should follow meteorology. If *Meteor*. IV is Aristotle's prolegomenon to his biological works, then it is not out of place at the end of the *Meteorologica*.'[27] Ammonius is alone in claiming that *Meteor*. 4 is both genuine, and in the right place.[28]

23. See 179,4-11 below.
24. *in Meteor*. 173,29f.
25. As cited by Ideler, *Meteor*. II, 379f.
26. Lee, ix.
27. Furley, 148.
28. in Olympiodorus *in Meteor*. 6,24-30.

Why the almost universal agreement that *Meteor*. 4, if genuine, does not belong with books 1-3? Coutant puts the argument succinctly:[29]

> The reasons for the attempts to move the fourth book from its conventional place are not far to seek. The first three books are concerned with meteorological phenomena as explained by the actions and affectivities of the moist and dry exhalations, and all such phenomena are ascribed to these agencies. Moreover, at the end of the third book a promise is made to furnish a study of the exhalations as they occur inside the earth. There two classes of matter are to be discussed, those bodies that are mined, and those that are quarried. But the promised discussion nowhere appears, either in the fourth book of the *Meteorologica* or anywhere else in the extant works of Aristotle.

All seem to be in agreement that the end of *Meteor*. 3 presents conclusive evidence that *Meteor*. 4 cannot be what follows. I disagree, and (siding with Ammonius) think that *Meteor*. 4 is just that, the final part of the treatise *Meteorologica*, written in partial fulfilment of the promissory note at the end of book 3. It is clear that the closing section of book 3 must be examined closely.

The section that concerns us runs from 378a12 to the end of the book at 378b6. This passage begins by summing up what preceded, an investigation into what are the products of 'separation', or 'exhalation' in the region (*topos*) above the ground. It goes on to claim that one must still discuss what is brought about by separation in parts enclosed in the ground. This precedes 378a16, which tells us what such products are. By means of two kinds of exhalations beneath the earth, one vaporous, the other smoky, two forms of body are produced, 'things quarried' (*orukta*), and 'metals' (*metalleuta* – Coutant's 'things mined'). We are told that the dry exhalation, by becoming fiery (*ekpurousa*), produces things quarried, examples being unmeltable (*atêkta*) rocks, and other unmeltables (I assume

---

29. Coutant, 8.

that the dry exhalation is here another name for the smoky exhalation).

Metals are said to result from vaporous exhalations, examples being iron, gold and copper. All are said to be liquefiable (*khuta*) or malleable (*elata*). Their creation is due to the compression (*sunthlibein*) and solidification (*pêgnusthai*) of the vaporous exhalation when it is trapped within rocks, brought about by the dryness of the rocks. It is then claimed that because of the manner of the creation of metals, they are in a sense water, but in a sense not. They could have turned into water (since they are a product of the compression and solidification of a vaporous exhalation), but instead of a liquid resulting, a solid is formed, via the compression of the exhalation (which, if not compressed, would normally condense into water).

The chapter ends by stating: 'Having spoken of all of these generally, we must now undertake to examine each kind separately.'[30] The kinds in question are things quarried and metals, not, as Coutant would have it, the exhalations themselves which form these. Aristotle promises a discussion of the products of these exhalations in the ground.

What precisely would one expect to follow all this? Well, one would want an account of things quarried, their coming to be and properties, and a similar account of metals. It is important to note that things quarried, a term which does not occur in *Meteor*. 4, are, at 378a22-4 characterised as those things which are unmeltable. Unmeltables do play a prominent role in *Meteor*. 4. In other words, a discussion of unmeltables is a discussion of things quarried, and is found in *Meteor*. 4.

In addition one would expect an account of compression, solidification, liquefaction and malleability. One might also expect an account of how metals are and are not water – an account of their elemental composition, and of how the action of heat might produce unmeltable things, and how dryness might yield solidification, and further information concerning both smoky and vaporous exhalations. One finds all of this in *Meteor*. 4, and so, contrary to received opinion, I take it that *Meteor*. 4 is correctly titled.

30. 378b5-6.

What does go missing in *Meteor*. 4 is an account of the properties of homogeneous things, in particular, unmeltables and metals, explicitly in terms of exhalations.[31] Yet why think that this is what one would expect to find in a fourth book of the *Meteor*.? More precisely, why not take it (*contra* Coutant and all others who think that *Meteor*. 3 promises a lengthy discussion of exhalations in the earth) that 378a16-b4 is the complete account of the formation of unmeltables and metals with respect to exhalations? I see no reason not to. *Meteor*. 1-3 has presented a rather exhaustive account of assorted phenomena with regard to exhalations. The end of book 3 tells us that the production of unmeltables and metals is like these other phenomena – via exhalations and their (by now) familiar behaviour – only underground. It makes perfect sense that the phenomena themselves (the production underground of things by exhalations) are not described in greater detail (as with the assorted descriptions of meteorological phenomena properly so-called) precisely because they happen underground. They are unobserved phenomena. In place of such a description of these phenomena, what one gets in book 4 is an explanation of such comings-to-be.[32]

One should also recall the beginning of *Meteor*. 1. There Aristotle tells us that, having already discussed the first causes of nature, natural motion, the four elements and their mutual transformations, it remains to discuss meteorology, the ac-

31. Although what is found is perfectly consistent with an account based on exhalations. For example, the claim made at 387b24 that certain bodies contain more earth than smoke is utterly baffling if not put in the context of enclosed exhalations as found in *Meteor*. 1-3, and 384b33 explicitly talks of the formation of metals via vaporous and smoky exhalations (the exhalations from water and earth respectively).

32. One might be worried by some terminological differences between the end of *Meteor*. 3, and *Meteor*. 4. In 3 metals are said to be meltable, the term being *khuta*, which is not found in 4. This is explained by the fact that *kheisthai* is a general term meaning 'to flow'. *Meteor*. 4 shows how 'flowing', or various things which we might think of as melting, is explained by the presence of water in the flowing thing, and so in *Meteor*. 4 one finds more technical terms employed like *hugrainesthai*. Only once flowing phenomena have been explained can these other, more technical, terms be introduced. A similar explanation accounts for the seeming importance of *sunthlibein* in 3, and its apparent unimportance in 4. Again *sunthlibein* is a general term for compression, or solidifying. In 4 Aristotle develops a whole typology of compression based on what is compressed, and how. The one occurrence of this term in 4 (384b10), like its occurrence in 3, relates 'compression' to the solidifying of what is compressed. Book 3's talk of dryness compressing and solidifying the vaporous exhalation is also consistent with 4, where watery things are said to be solidified by dry heat (4.6), and the dry individually is said to cause solidification (384b4).

count of regular natural phenomena near the earth and in the earth. Such phenomena are said to be affections common to both air and water, fire, and earth and its parts. After this, Aristotle states that one must give an account of animals and plants. *Meteor.* 1.2, after opening by reminding us of the natural motions of each of the four elements, states that 'the subject of our inquiry, as we say, is the accidental affections (*ta sumbainonta pathê*) [of the four elements]',[33] which all sublunary things are composed of. This, I take it, promises a discussion of the characteristics of the four elements which considers characteristics other than the contraries they are characterised by. We know from *GC* that fire is hot-dry, air hot-moist, water cold-moist and earth cold-dry, and we know some of the properties the elements have directly in virtue of the contraries they are characterised by (fire sorts like to like, for that is what the hot does). We are here led to expect a discussion of other characteristics of the elements, which is found in book 4.

In particular, a perfect candidate for what the accidental affections of the elements might be is precisely those properties they contribute to the composites which they compose. For example, earth is not essentially able to be softened by heat, but bodies composed mainly of earth are, and so an accidental affection of earth is to cause bodies composed (mainly) of it to be softened by heat.

Meteor. 4 seems to be just what one would expect, given the beginning of *Meteor.* 1, and the end of *Meteor.* 3. From the beginning of *Meteor.* 1 one would like a work discussing varied characteristics of the elements, and in particular non-essential characteristics of them. This is just what *Meteor.* 4 is all about. One would also like a work which might make a natural bridge to a discussion of biological phenomena. As Furley and others have noted, *Meteor.* 4 also plays this role. In fact, *Meteor.* 4 is carefully crafted to fulfil the promissory notes of both book 1 and 3. The discussions of unmeltables and metals (fulfilling the promise of *Meteor.* 3) is expanded to include a discussion of many (perhaps all) the most basic properties of homogeneous stuffs (fulfilling the claim in *Meteor.* 1 that one must discuss what happens naturally to material things). These properties

33. 339a20-1.

are described in so far as they are products of the powers, both active and passive, of the four elements (so fulfilling the promise of *Meteor*. 1 to discuss the accidental affections of the elements). And, the discussion of homogeneous stuffs in *Meteor*. 4 often includes organic homogeneous stuffs, while often mentioning how heterogeneous organic things differ from homogeneous ones in origin and function, and ends by promising a discussion of anomeomerous things,[34] 'people, plants, and the like', thereby fulfilling yet another promise made in *Meteor*. 1. Aristotle realises brilliantly that a discussion of homogeneous stuffs naturally leads to biology, while at the same time forming a perfect culmination of a discussion of *Meteorology*, the study of 'whatever happens according to nature to bodies',[35] in the sky, water or earth. I therefore agree with Furley that '*Meteor*. IV is Aristotle's prolegomenon to his biological works', and that 'it is not out of place at the end of the *Meteorologica*', but, unlike Furley, think that it belongs at the end of the *Meteor*. as the last part of this work. *Meteor*. 4 is both authentic, and properly placed.[36]

Establishing the authenticity and placement of *Meteor*. 4 does not, in and of itself, prove its importance. Yet in Aristotle's extended discussion of the composition of the natural world *Meteor*. holds the conceptual 'middle ground' between *GC* and the biological corpus, and this already suggests that it may be an important work. *GC* discusses the four elements, their properties and interactions, *Meteor*. 4 discusses the properties of homeomers composed out of the four elements and their interactions, while the biological corpus studies wholly formed living beings, themselves composed out of homeomers.

---

34. Things whose parts are unlike the whole. For example, parts of a face are not faces. For a discussion of these terms, see note 17 to the translation.
35. 338a26-b1.
36. By the above discussion I do not intend to support any particular position concerning the original organisation and division of the works of Aristotle. Whether *Meteor*. was actually composed in four books, or whether these books were meant to be separate, yet related, treatises, or what we should take a book to be, is not something I wish to take a stand on. The question of placement, and how my view differs from that of Furley's, is this: I think that the work which we have as *Meteor*. 4 is that work promised both at the end of what we have as *Meteor*. 3, and near the beginning of what we have as *Meteor*. 1. In *this* sense it follows *Meteor*. 1-3. The question of what forms a unified work prior to printing, and when one is faced with, as we are, 'school notes' (in some sense or other) is a difficult one. I thank S. Menn for urging me to clarify my position on this.

These discussions use, modify, and sometimes supersede, much of the natural philosophy developed in works such as *Cat.* and *Phys.* In particular, *Meteor.* 4's theory of the elements and the contrarieties is bold, central to Aristotle's analysis of alteration and substantial coming to be, revealing of much that is of interest in his theory of matter, perhaps revisionary, and certainly understudied. In what follows I will sketch via Alexander's commentary some of the more interesting aspects of the theory concerning the relationship of the contraries to the elements. In the end I am sympathetic with many of Alexander's interpretations of *Meteor.* 4. I hope at least to hint at much of what is important in *Meteor.* 4 itself, and suggest questions worth pursuing concerning how the doctrines found in *Meteor.* 4 have an impact on assorted topics central to Aristotelian natural philosophy.

3. The elements and the contraries, matter and change

I believe that *Meteor.* 4 develops a theory hinted at in *GC*, namely that the four elements (fire, air, water and earth) are *composed* out of the four contraries (the hot, the cold, the moist and the dry). In addition, I believe something closely related to this claim, that the elements are hylomorphic. In other words, there is that which is the matter of the elements, and that which is their form. I do not think that the elements themselves are 'simple' matter, or 'pure' matter, or fall in any way short of being compounds (although they may be the most simple *bodies*, or bodies that are not themselves compounded out of bodies). I think that *Meteor.* 4 is the decisive (although not the only) text for determining these issues. I also believe that it is useful to couple the account Aristotle offers in *Meteor.* 4 and elsewhere of the relationship of the contraries to the elements with his theory of alteration and change, found primarily in *Phys.* and *GC*. Together these suggest a solution to the Aristotelian problem of change (to be discussed below) different from those hitherto advanced.[37]

37. Some of what follows can be found in my Ph.D. thesis, *Body, Matter and Mixture: The Metaphysical Foundations of Ancient Chemistry*, University of Illinois at Chicago

The theory of matter advanced in *Meteor*. 4, and commented on by Alexander, is one that has gone largely unnoticed by modern commentators. One of the most recent studies to examine at length Aristotle's theory of matter is M.L. Gill's *Aristotle on Substance: The Paradox of Unity*. One finds here a sustained, and often penetrating, discussion of Aristotle's theory of matter, particularly concerning the status of the elements and contraries. By taking Gill's account as 'state of the art', I hope to highlight the sense in which *Meteor*. 4's treatment of the status of the elements and the contraries has been undervalued. For while I find much to agree with in Gill's account, I disagree with her interpretation of the relationship of the elements to the contraries. The source of my disagreement is, to a large part, based on texts found in *Meteor*. 4.

The following two passages from Gill nicely sum up her position (which I disagree with):[38]

> ... I, too, envisage a cosmic scale, at one end of which is pure form and at the other pure matter. But contrary to the traditional conception, the cosmic scheme is grounded not in an indeterminate prime matter but instead in a set of simple and identifiable elements .... Unlike other higher materials, which are composites of a simpler matter and form, the elements, though they can be analysed into matter and form, are, ontologically speaking, pure matter: they are not composites of simpler ingredients organized by form. I suggest that ... Aristotle calls the elements matter in the 'strictest' sense because they are the ultimate matter in the sublunary sphere.

Circle, submitted 1988. I am in sympathy with much that M.L. Gill has to say in her important *Aristotle on Substance*, Princeton 1989, although we disagree rather fundamentally on how the relationship between the elements and the contraries ought to be viewed.

38. Gill's account is particularly appropriate to serve as a benchmark for an examination of *Meteor*. 4, since many of those who disagree with her, disagree only with respect to whether prime matter need be countenanced as playing the role of pure matter. In other words, they agree with that aspect of her theory which I disagree with, namely her account of the relationship of the elements to the contraries. I believe that both Gill's account, and mine, need not rely on prime matter, and so are evidence that Aristotle had no such notion. However, even if I (we) are wrong about this, grafting prime matter onto my reconstruction of Aristotle's theory leaves intact its differences from Gill. I will not here directly discuss issues surrounding the putative existence of prime matter.

and:

> ... Aristotle treats the elements as the ultimate ingredients out of which all bodies of greater complexity in the sublunary realm are composed. We are not entitled to ask out of what further ingredients the elements themselves are composed because there are no simpler ingredients. Unlike higher material bodies, which are composed out of the elements and so can be analysed into the elements as their matter, the elements themselves are not composed out of any simpler stuffs and so cannot be similarly analysed .... Although the elements are not composed out of simpler matter, they can nonetheless be analysed to determine what survives an elemental transformation and what is replaced. So one can specify a factor corresponding to the form and a factor corresponding to the matter of an element. But elemental matter is not an ingredient, and elemental form is not an arrangement or structure imposed on an ingredient ....

These two passages summarise a position I wish to call into question. This position holds that:

(1) The lowest level of composition is that of things composed out of the four elements.

(2) The four elements themselves are simple. They are not composed out of anything further. They are without ingredients.

(3) In virtue of this, the four elements are 'ontologically basic' matter, or matter 'strictly so called'.

I disagree with each of these claims, and believe that the elements are composed out of simpler ingredients, and are not simply analysed into something 'corresponding' to matter and something 'corresponding' to form. My reasons for thinking this are both textual and philosophical. I think that there is textual evidence that Aristotle takes the elements to be composites, and that this is to be expected, since, as I shall show, he clearly has philosophical reasons for thinking this.

In a nutshell, I believe that for Aristotle, the elements are composed out of the contraries that characterise them. Fire is composed out of the dry and the hot, air out of the hot and the

moist, earth out of the cold and the dry, and water out of the
moist and the cold. The elements may be the simplest *bodies*
in the Aristotelian universe, but they are not simple *tout court*.

The dual notions of matter and form are slippery enough to
make it difficult at times to pin-point the differences between
competing positions concerning what is matter and what is
form. This is partly because the same entity can, at one and the
same time, play the role both of the matter of something, and
of the form of something. Most things can be understood as both
form and matter, or more properly, as both the form of something and the matter of something else. Gill suggests that
something is *truly* matter when it is an ingredient of that of
which it is the matter, while something is *truly* form when it is
an arrangement or structure of 'true' matter. Yet it is unclear
what being an ingredient amounts to. This gets to the crux of
the problem. I contend that on any account of 'being an ingredient' that one might plausibly derive from the Aristotelian
corpus, the contraries count as ingredients of the elements.[39]

One test which many, including Gill, treat as persuasive is
this: in order for some X to be an ingredient in some Y, X must
pre-exist separately from Y, and there must be a temporal act
of the composing of Y out of X. The bronze of a bronze statue
pre-exists the statue, and there is an act of the creation of the
statue out of the bronze as an ingredient. Yet such an account
is problematic, for it becomes unclear whether the elements
count as ingredients, or true matter, given this test. The reason
is that Aristotle is committed to the claim that there are no
instances of pure elements in the sublunary world. *GC* 2.8 is
devoted exclusively to arguing this point. One never has pure
earth, air, fire or water pre-existing, and then some process of
their being formed into bronze. Yet surely the elements are the
true matter and the ingredients of bronze.[40]

One might claim that the elements, like the bronze, need not

39. From the above quotes it should be clear that Gill's account hinges on the
distinction between being such to be *analysed* into matter and form, and being
composed out of matter and form. In what follows I will argue that no such plausible
distinction can be drawn, at least none that can do the work Gill requires it to do.

40. See also *Metaph*. 1040b5f., which states that the four elements are always
potential, and never actual. One may have the elements pre-existing *in potentiality*
that which is composed out of them, but this is no help, since the composite itself also
exists in potentiality prior to its composition.

pre-exist that of which they are the matter as isolated, free-floating entities. The bronze of the statue *was* the bronze of a bar of bronze, or of a pile of bronze. *Something* brazen pre-existed the statue. But what pre-exists the bronze which is earthy or watery or airy or fiery? A pile of bronze is actually a pile *of* bronze, but the four elements seem only to exist *potentially* in compounds. We need to offer an account of such existence in potentiality so that only the *elements* come out as pre-existing in the appropriate sense, and not, say, everything in the universe. I will discuss below how we ought to understand the existence in potentiality of the elements in compounds.

This pre-existence test fails on other grounds. For Aristotle the universe is uncreated. Whatever Plato may have held, Aristotle does not think that the present world order was created out of pre-existing disorderly matter. In other words, there may very well be, at the centre of the earth perhaps, a rock that was never created, but has always existed. Yet should we not still be able to claim that it contains the four elements as its matter? No, according to the temporal test, since there was no temporal act in which this rock come to be out of the four elements (or out of anything else). For these reasons, the temporal test does not seem satisfactory.[41]

What tests does Aristotle employ for being an ingredient of some compound? It seems that things are *composed* out of their ingredients (regardless of whether there is a temporal act of their composition), and so an examination of passages on composition may shed light on what it is to be an ingredient. There are, scattered throughout the corpus, texts discussing various criteria for composition. One which Gill discusses is *GA* 724a20-30. The relevant passage is:

> But there are many ways in which something can come to be out of (*ek*) something else ... we say that a statue comes to be out of (*ek*) bronze and a bed out of wood, and so on in all the other cases where we say that the thing coming to be comes to be out of (*ek*) matter, the whole being [composed] out of (*ek*) something inhering (*enuparkhontos*), which is put into shape.

41. *DA* 418b9 discusses 'the eternal upper body', namely aether. By the temporal priority test we could not say that there is that which is the matter of aether.

This passage suggests a number of points. First, it suggests that there is a sense in which things are out of (*ek*) their matter, the sense being that they are *composed* out of their matter. This passage is part of a longer discussion of the ways in which something can be 'out of' something else. Being composed of this something else is but one sense, clearly an important one. So, from this passage we learn that there is a compositional use of '*ek*'. In the translation that follows I will translate compositional *ek*s by 'out of'. Now, if things are composed out of matter, and if in general the things that something is composed out of are the ingredients of that thing, matter is an ingredient of the things composed out of matter.

This passage also suggests the following test for being the matter or an ingredient of something: if X is the matter of Y then X inheres in Y.[42] This entails that the matter out of which something is composed does not pass away when it becomes the matter of such a thing, for it inheres in the thing. The casting of a bronze statue is the coming to be of a statue, but it is not also the destruction of bronze (although it may be the destruction of a bar of bronze). Aristotle claims that the contraries inhere in the elements, and so by this test they may be the matter of the elements (*GC* 331b4f.).[43] In fact, Aristotle states in the same passage that the elements can come to be from each other, and such coming to be is easy, precisely because 'the couplings [that is, the contrarieties] inhere in (*enuparkhein*) the adjacent elements' (*GC* 331b4).[44]

A passage from the *PA* explicitly recognises that the contraries are the ingredients of things, which entails that the elements, and so all composite bodies, are composed out of them (*PA* 646a13-18):

42. I word the principle this way (as opposed to 'If X inheres in Y, then X is the matter of Y') since this is not a sufficient condition for being the matter of something.

43. Alexander also talks about the contraries being inherent in that which they qualify (189,13; 192,10,13; 202,30). There is further textual support in Aristotle for the existence of the 'compositional *ek*'. Aristotle devotes a chapter in his 'philosophical lexicon' (*Metaph.* 5.24) to the various meanings of 'to come from (*ek*) something'. Here we read: 'To come from (*ek*) something means, 1. to come from something as (*hôs*) matter, and this in two ways, either according to (*kata*) the first genus, or according to the last species, for example, on the one hand all meltable things are out of water, but on the other the statue is from bronze' (1023a26-9).

44. Aristotle is also clear that during elemental transformations, some contraries remain, and so, presumably, inhere as matter in the new element that comes to be.

First one may assume the composition (*sunthesis*) out of what some call the elements, earth, air, water and fire. Yet it is perhaps more accurate to say composition out of the powers (*dunameôn*), and not from all of these, but as I said previously elsewhere,[45] the moist and dry and hot and cold are the matter of composite bodies.

It seems apparent that the whole of *Meteor.* 4 is in agreement with this passage. For instance, we read:

Unqualified natural becoming is a change brought about by (*hupo*) these powers [the hot and the cold], when they are in a certain ratio, out of (*ek*) the underlying matter of each natural thing, this [i.e. the underlying matter] being the aforementioned passive powers [the moist and the dry]. (*Meteor* 4.1, 378b33-79a1)

In addition, Alexander is in agreement with this view of the composite nature of the elements, both in his commentary on *Meteor.* 4, and in his *Quaestiones*. For example, we read:

For it is least of all clear what the particular purpose is of each of these bodies [i.e. the elements], in which matter plays the greatest part, that is, which are nearer to matter. Those are nearest to matter which are the first to be produced out of matter, and these are the elements. Hence in compounds these have the definition of matter ... (*in Meteor.* 4, 224,12-16).

The elements are not mere matter, but they are the nearest to matter. They are the first things out of matter, their matter being the contraries.

Let us take stock. I claim that the elements are composed out of the four contraries, that the contraries are ingredients of the elements. The elements themselves are not simple matter, or merely matter, for there is that which is the matter of them. They are, however, the simplest *bodies*, which Aristotle

---

45. This is often taken to be a reference to *GC* 2.2, but it may be a reference to *Meteor.* 4.

repeatedly calls them. The test for being the matter of something is being that which something is composed out of. If the
elements are composed out of the contraries, then the contraries are the matter of the elements. This is no mere 'analysis' of
the contraries as matter. The contraries are the matter of the
elements in as robust a sense of matter as one could wish.

4. Elemental hylomorphism and the contraries

I claim that for Aristotle the elements are hylomorphic, composites of form and matter. We have already looked at some of
the evidence that there is matter of the elements. Yet what of
their form? Is there any evidence that there is 'an arrangement
or structure imposed' on the matter of the elements, which
could be the form of them? I think there is. (We shall see that
Alexander is certain that there is.)

Meteor. 4 develops a theory of the elements which has as an
essential component the claim that the two active contraries,
the hot and the cold, normally act as the formative agents on
the passive two, the dry and the moist. The active contraries
are the efficient causes of the formation of the passive contraries. When, for example, the dry is informed by the hot, what
results is the coming to be of fire. It is, of course, the fire that
is the proper subject of the form of fire. Since each element is
composed of one active contrary and one passive contrary, each
element is composed of an informed contrary, and an informing
contrary.

This is not to deny that the elements are composed out of all
four contraries as matter. It is just that the active contraries,
in so far as they are active, are the efficient causes of the
formation of the passive contraries. There is nothing improper,
given Aristotle's views, in something acting as an efficient
cause in one respect, and a material cause in another. The
active contraries, in so far as they compose the elements, are
material causes of the elements. In so far as they are responsible for the formation of the passive elements they are efficient
causes of the elements' coming to be.[46]

---

46. It is instructive to read *Meteor.* 4.12, which claims that the simpler something is,
the more difficult it is to differentiate the four causes associated with that thing.

Failure to recognise this might tempt one to see a potentially problematic disagreement between *Meteor.* 4 and *GC.* Throughout *GC* (as we shall see) all four contraries are conceived of as the matter of the elements. On the other hand, *Meteor.* 4 often reads as if only the passive two contraries are the matter of the elements. This is, however, only a superficial disagreement. *GC* is concerned primarily with the composition of the elements and with the contraries' role as the material substrate through elemental transformations. It therefore emphasises the role all four contraries play as the matter of the elements. *Meteor.* 4 is mainly concerned with explaining the properties of the elements, and of homeomers, and so stresses the informing powers of the active contraries. As we shall see, particular properties of homeomers are explained by means of the different ways in which the active contraries can inform the passive contraries.

Let us now examine some texts. We are told at *Meteor.* 379a1 that the hot and the cold produce change in bodies by mastering the matter, that is the passive contraries. As we shall see in the next section, when matter is mastered, it is organised or informed (*sunistanai*), and in general becomes more complex.[47] An equivalent claim is made at 380a7f., where concoction is said to result from inherent heat mastering indeterminate matter. Again, as we shall see, concocting is the imposition of organisation or form. Spontaneous growth of living organisms in decayed matter is said to be due to natural heat organising matter (379b6f.). Rawness is also said to result when matter is not mastered by heat (380a26f.), while the solidification of compounds of earth and water is due to the dry being organised by the cold via the departure of heat. All this suggests that the active contraries inform the passive contraries by imposing organisation on them. This informing action of the hot and the cold is made explicit at 384b24 which states 'that it is clear that bodies are organised (*sunistatai*) by hot and cold'.

An interesting passage in 4.10 (388a17f.) tells us that the material cause of the organisation of homogeneous bodies is the dry and the moist, while the efficient cause of this organisation

47. I will discuss below the range of meanings of *sunistanai*. For now I will translate it by 'organise' and its cognates. I mean 'organise' in its widest possible sense, certainly not limited to the spatial organisation of parts.

is the hot and the cold. Later in 4.10 we are told that one can distinguish whether a thing is of earth, of water, or of both, by whether it is organised by fire, or cold, or both (fire here standing in for the hot). A passage in 4.11 confirms this, telling us that things composed out of more than one element have heat, for they are organised by concoction brought about by heat (389b7f.). As a last bit of evidence, an important summarising paragraph in 4.12 informs us that heat and cold and the motions they produce are sufficient for producing all the homeomers (390b2f.).

This model, which has it that the active contraries inform, or organise, the passive contraries, is clearly one which Alexander endorses. He uses the term '*eidopoiein*' (to inform) to describe what it is that the contraries do to the elements (179 etc.). More specifically, it is the active contraries which inform the passive contraries as matter (181 etc.). Concoction, for instance, is characterised as an organising resulting from the hot mastering the moist (188 etc.), while rawness (an inconcoction) results from heat's failure to organise something (189-90). In general, Alexander characterises concoction as the change of a substrate via the acquiring of a certain form (193).

That the contraries' action on one another imparts organisation is clear. Alexander tells us that the moist causes the dry to be organised and not crumble, while the dry causes the moist not to dissipate or scatter, and organises it.

The close relationship between organising and informing should be clear. Alexander calls affections 'forms coming to be in matter' (201), and claims that 'the affection produced in the subject is produced because of the presence of hot or cold' (202). It is therefore difficult not to conclude that the hot and the cold inform that which they are present in. Again, the opening line of Alexander's commentary on 4.8 states that:

It is clear from what has been said ... that composition (cognate of *sunistatai*) occurs in bodies by means of hot and cold. For the hot and cold, as has been said, produce the composition of bodies either by drying them or by thickening and solidifying them. This is because these are the agents of composition, in all bodies which are both mixed and have composition, that is, determined [bodies].

Alexander expands upon this in his commentary on 4.10, where he tells us that the matter of the homeomers is the dry and the moist, the efficient cause being the hot and the cold, 'for these bring about the composition of homeomerous [bodies] out of those' (219).

An interesting passage in *Meteor.* 4.12, along with Alexander's commentary on it, deserves separate treatment. I read it as arguing that the elements are not pure matter, and so must be hylomorphic. The text reads as follows (390a3-9):

> For 'the for the sake of which' [i.e. the final cause] is least clear wherever matter predominates. For, taking the extremes, matter is nothing other than itself, and essence is not other than definition. The things in between are analogous to that to which they are nearer, seeing that each of these is for the sake of something ....

This passage is embedded in a discussion of the differences between the form and the matter of various composites. Aristotle claims that the final cause of things is other than their matter. He insists that all the 'works of nature' have a final cause, but that this final cause or function is harder to discern, the simpler the thing considered should be. The function of animals and parts of animals, or anomeomers in general, is often easy to discern, while that of homeomers is more difficult.

Aristotle goes on to state explicitly that this function is even less clear in the case of the elements: 'Similarly with fire, but it is less clear than with, e.g., the tongue, what its natural function (*ergon*) is' (390a15-16). Yet the fact that it is less clear in the case of the elements entails that they *do* have a function distinct from their matter, and so they have a form.[48] They cannot be 'pure matter'. These passages make it clear that the elements are being held in opposition to 'pure matter'. True, the elements are closer to pure matter than to form, but they are not pure matter. The elements have functions, and so form. This discussion strongly suggests that here at least Aristotle views the elements as hylomorphic compounds of form and matter.

---

48. In general Aristotle thinks that whatever has a function (*ergon*), has a form.

Alexander's insightful comments on this passage are most helpful, and support the interpretation just presented. We read:

For it is least of all clear what the particular purpose is of each of these bodies, in which matter plays the greatest part, that is, which are nearest to matter. Those are the nearest to matter [224,15] which are the first to be produced out of matter, and these are the elements. Hence in compounds these have the definition of matter .... [49] Those things are [nearer] to matter which have underlying matter separate from forms.[50] For such is the [material] substrate (*hupokeimenon*) of the elements, since each of [224,30] the simple primary bodies, he says, is for the sake of something (*estin heneka tou*), and since each of these [the four elements] is not [merely] water or earth taken in no matter what condition, nor something [else, that is, neither air nor fire], just as neither flesh nor liver [is flesh or liver taken in no matter what condition]. But just as [flesh and liver are flesh and liver] when they produce that which is expected from them (*hotan to autôn parekhetai*), so too [the elements are elements], when each of them retains its own nature and its particular function (*ergon*).[51]

Although the above passage in Aristotle is not without its mysteries (For instance, in what sense is form equated with definition?), both Aristotle and Alexander are emphatic that the elements have both form and matter, and so have functions.

---

49. In other words, the elements are the matter of compounds. Note that Alexander does not consider the elements themselves to be 'pure' matter, only to be closest to matter.

50. Meaning, I take it, those which have underlying matter itself without form – matter which is not itself hylomorphic.

51. In other words, just as a liver is properly a liver when it is a functioning liver, but not when served up with fried onions, so too with the elements. Every element has a function which it must be able to perform (be actually performing?) in order to be properly, and not equivocally, the element that it is.

5. *Sunistanai, horizein,* and the actions of the contraries

Following Gill and others, I have couched talk of form in terms of imposed arrangements or organisation. The term 'organisation' has no obvious correlate in Greek. Yet if the form of some X is that which organises X, or is the organisation itself, it is important to have a firm grasp on the concept of organising. In fact, *Meteor.* 4 is concerned with organisation as much as it is concerned with anything. *Meteor.* 4 is about the various ways in which organisation can be imposed on matter, specifically the various ways in which the dry and the moist as matter can be organised by the hot and the cold. Given that being organised in a particular way is so closely related to the having of a particular form,[52] it should come as no surprise that the concepts of form, organisation, determination and definition are interrelated in *Meteor.* 4. In what follows I shall sketch an account of these interrelations, particularly as Alexander conceives of them.

The terms Aristotle and Alexander often use to talk about the organisation or informing of something are cognates of *sunistanai. Sunistanai* in *Meteor.* 4 is translated by Lee as 'to compound', 'to constitute', 'to acquire consistency', 'to form', 'to condense', 'to pack closer', 'to compose' or 'to produce'.[53] Coutant preferred cognates of 'condense'. It is impossible to translate this term the same way throughout Alexander's commentary, if any sense is to be retained. I often use 'compose', a fairly neutral term, yet 'condense' is sometimes clearly preferable. Sometimes 'bring into being' seems best to capture its meaning. I consider all these meanings to fall under the umbrella concept of 'organisation'. It should therefore be clear that I intend 'organisation' to be taken in a very general sense, one which is certainly not limited to the spatial rearrangement of parts.

Aside from the multiplicity of meanings of *sunistanai* and its cognates, the translation and interpretation of this term is complicated by a curious feature of its use. Sometimes it is used

52. Since, as we shall see, the manner in which something is organised affects the functions it can perform.
53. *Sustasis,* which figures in Alexander's commentary, is not found in *Meteor.* 4.

to talk about the organisation, or composition, which something has. When it is used this way, we find locutions such as:
'X is composed (cognate of *sunistanai*) *out of* Y and Z', where it
is clear that Y and Z are the material constituents of X. Other
times it is used to talk about the means by which something is
organised or composed, the cause of a given composition. In
these cases one finds locutions such as: 'X is composed (or
organised) *by* Y and Z', where it is clear that Y and Z are the
efficient causes of X's having organisation or composition.

Without wanting to make too much of the point, these two
uses of *sunistanai* come close to picking out matter and form,
respectively. When Alexander states that something is composed from or out of something else, he means that this thing
has that which it is composed from, or out of, as its matter, it
is a particular composite of this matter.[54] On the other hand,
when Alexander states that something is composed, he often
means that the matter of the thing is arranged in a new, and
usually more complex, way. Often we find 'composed by' when
Alexander has this meaning in mind. Coming to have a composition is in this sense, at first, coming to have definite bounda-
ries and a fixed internal consistency, that is, becoming 'formed'.

Alexander's commentary suggests a very close relationship
between the notions of composing (that is, taking on a new
organisation or composition), defining and determining. When
something is determined, it takes on a new, and usually more
complex, composition. The determining of something is the
imparting of composition or organisation. In addition, in order
for something to be most fully what it is, it must be determined,
or be determinedly what it is. When something is determined
and given the composition or organisation most proper to it, it
is what it is most fully. An unripe apple is not 'fully' an apple,
it is an immature apple, and upon ripening, which is a determining, it will become a mature, fully formed apple.

This being most fully, or properly, what one is, is the being
of something according to its definition. *Horizein* means both
to determine and to define, and in *Meteor.* 4.12 the close
relationship between these two notions makes it difficult to

---

54. In what follows, I will concentrate on the meaning of *sunistanai* related to
informing. For 'composed out of' (as matter) see, esp., *in Meteor.* 4.8 below, where, for
instance, metals are said to be composed out of water and earth.

decide how to render the term. I have often translated it as 'define/determine', since substituting either of the pair often renders the sentence intelligible.

First, I will discuss the relationship between determining and *sunistanai*, which is a concern in the body of *Meteor.* 4 , and then move on to a discussion of the dual meanings of *horizein*, as found primarily in *Meteor.* 4.12.

We are told by Alexander early on in his commentary that Aristotle considers determining to be a concentrating and thickening, or a solidifying, 'so that what previously was without its own shape by reason of moisture and always took its shape from the surroundings assumes a shape of its own and has it own limit' (180,12f.). Interestingly enough, it is often the dry and the moist which are said to be determined. This is yet further evidence that they are material, since it is otherwise hard to see how they can be thickened, concentrated or solidified. In fact, Alexander claims that the dry and the moist are defined 'by the words "easy to determine" and "hard to determine" ' (181,12). I will discuss below precisely what the four contraries are, given my reconstruction of Aristotle's theory.

In general, Alexander's *in Meteor.* 4.1 argues that determining is what accompanies natural generation, and that this determining is often accomplished by means of a concoction. 'For [every compound composed by nature] is generated in the beginning from the moist and the dry; since the dry is determined by the moist and takes on a boundary and shape of its own because of its mixture with this [that is the moist]; this is brought about by the active powers, which are heat and coldness' (183,9-12). Spontaneous generation is said to result from the natural heat of putrefying things organising the moist matter of that which is putrefying. This is, I take it, a determining of the moist matter. Later, Alexander clearly (and correctly) thinks that concocting is equivalent to determining, since his comments on 380a20 ('in that which is determined by natural heat and coldness') assume that determining by heat and coldness is concocting by heat or coldness (189,5f.). In fact, Alexander goes on to claim that 'the heat determines the moist and the cold as matter by mastering [them] in concoction' (189,9-10). This is an explicit statement of what Alexander takes concoction to be, namely, a determining. So, any evidence that Alexander takes concocting to be organising is at the same time evidence that he takes determining to be organising, and there is ample such evidence.

For instance, in Alexander's account of ripening, a type of concoction, he claims that ripening involves an organising and thickening of air or water (189,16).[55] In fact, Alexander tells us that 'everything which ripens is organised' (190,5). Rawness, the inconcoction parallel to ripeness, is said to result when things 'have not been mastered by the proper heat nor organised' (190,28). More specifically, 'whatever is capable of changing and being organised by heat, yet does not organise, nor is affected, is called raw' (190,30).

The relationship between concoction and determination is again made explicit in Alexander's account of boiling (a type of concoction), which results when 'the moist in it [the boiling thing], being undetermined, is both concocted and (*te kai*) determined by the heat in the external moist' (191,23-4). A similar account is given of the relationship of the dry to the moist. Alexander, following Aristotle, characterises the moist as being easily determined, and the dry as being hard to determine. However, 'the dry, by the mixture of the moist with it, acquires the ability to be determined ... for the moist causes the dry to be organised (here = congeal), and not crumble, and the dry likewise causes the moist not to dissipate nor to scatter, but to be organised (= to have consistency)' (198,35f.). It is for this reason that Alexander goes on to claim that 'every body both organised (= composed) and (*te kai*) determined is [composed] out of both moist and dry because each of these are undetermined if taken unmixed' (199,7-9).[56]

Determining and organising are often said to be processes in which something becomes (more) solid, or thicker.[57] Thickening is said to occur when the moist in the thickening thing departs,

---

55. See also 204,15f., where liquefaction is said sometimes to result from the change of an organised (here meaning condensed) exhalation, for instance, pneuma changing into water. 208,28 has the air in olive oil being both organised (= condensed) and thickened in its change into water due to the cold, while 211,1 claims that the earth in milk 'is both organised and (*te kai*) thickened and solidified'.

56. Alexander uses the locution 'both organised and (*te kai*) determined', or 'both determined and (*te kai*) organised' often, stressing the close relationship between these notions.

57. 209,26 claims that solidifying is a thickening.

and the dry in it is organised (= condensed or compacted). In fact, organising often (if not always) seems to entail the thickening and solidifying of that which is organised, which explains why the best translation of *sunistanai* is often 'to condense'. In general, the move from air to water to earth is a transition towards things with greater organisation (it is clearly a move from the less to the more dense). Indeed, 211,1 claims that the earthy part of milk, which becomes whey, 'is organised (= condensed) and both thickened and (*te kai*) solidified', making clear the close relationship between these processes.[58] In fact, Alexander supplies us with a definition of solidification that he thinks Aristotle would agree to, namely that 'solidification is an organising (= condensing) of water, or earth and water ...' (212,29).[59] After summing up what Alexander sees as the close interrelationship between organising, thickening and solidifying, he states that the hot and the cold 'are the agents of the organising of bodies, in all bodies which are both mixed and (so) have organisation, *these being determined* [*bodies*]' (213,2f., my emphasis).

We have now examined the close relationship between the organising and the determining of something. When something is determined it is organised. This organising, and so determining, is often a thickening, or a solidifying, of that which is being determined. In one sense, things being determined take on definite spatial limits of their own. In this sense they are organised, and one can see why thickening and solidifying would accompany such determining. Yet in another sense, things being determined achieve full actualisation. That is, things when determined are what they are most fully. It is in this sense that things when determined take on the organisation most appropriate to what they are. Determining is therefore also 'form-giving', the term *eidopoieisthai* which Alexander is so fond of. So, when something is determined it takes on a particular shape and/or form. Yet we also know that the term *horizein* means 'to define' in addition to 'to determine'. *Meteor.*

58. See also 211,21, when the blood of the stag is said not to 'solidify and (*te kai*) condense'.

59. 213,1 expands on this idea, claiming that the hot and the cold 'produce the organisation of bodies either by drying or by thickening and solidifying them'. See also 220,14, where organised things are called 'solidified and hard things'.

4.12, and Alexander's commentary on it, investigate the relationship between determining and defining.

Alexander tells us that the purpose of *Meteor*. 4.12 is to tell us what each homeomer is, 'not according to the substrate (for he has said this) but according to the definition (*logos*)' (223,8). This 'being according to definition' is said to be what something is in form, which is explicitly identified with the capabilities (*dunasthai*) something has (223,25). It is for this reason that, for example, a corpse is only equivocally a person, 'for no one defines a dead person as a rational mortal animal' (223,29). In general 'things which are for a certain purpose, when they are no longer able to fulfil this purpose and retain only their shape, are said [to be what they are] equivocally' (224,6f.). Yet, as both Alexander and Aristotle realise, the simpler something is, the more difficult it is to discern its function. It is most difficult to discern the function of the elements, the simple bodies, 'things which are near to matter'. Yet again, both are clear that 'the simple primary bodies' are 'for the sake of something'.

We can already guess at some of the interrelations developed in *in Meteor*. 4.12 between the various concepts that we have been investigating. The definition of something, or 'being according to definition', will involve a capacity or capacities which something must have in order to be what it is. To have this capacity, or these capacities, is to exemplify the form of the sort of thing something is. We have seen that the determination of something is an informing of it by means of an organising of it. New organisations allow something to have new capacities. The alteration of the organisation of something will therefore alter its capacities. In other words, if to gain organisation is, among other things, to gain capacities, and there is an organisation (and so a set of capacities) which something has when it is most fully what it is, then it is fair to say that something is determined by having (or gaining) a certain capacity, and is therefore also defined by this capacity. To be determined would be to take on an organisation such that one gains a characteristic capacity or set of capacities. To have these capacities would be to exemplify the form of this thing, to be this thing according to its definition.

It will become clear from the following translation (and from *Meteor*. 4 itself) that Aristotle, and so too Alexander, is interested in organisation primarily with respect to the capacities things have because they have been organised in particular ways. *Meteor.* 4.8 discusses the different passive capacities things have because of their composition, or organisation. Let us now look at those passages in Alexander's commentary where *horizein* seems to mean simultaneously 'determine' and 'define'.

We are told (225,5-8) that the substance (which here is synonymous with the being) of both anomeomers and homeomers is defined/determined according to their 'for sake of which' or final cause, through which such things are given form. This 'for the sake of which' is said to be a function. Now these claims make sense whether one is speaking of defining or determining. Something is determined when it takes on a particular organisation which gives it particular functions. This taking on of a new organisation, or determining, will often come about by means of concoction, thickening and solidifying. Similarly something is defined, or exists according to its definition, when it has a particular function because it has a particular organisation. 'Those that can perform the function proper to them are truly that which they are called' (225,6-7), and are thereby fully determined and are what they are according to their definition, and not merely equivocally. These 'natural functions' are said to be unclear in the case of the simple bodies (the elements), and so it is unclear what defines/determines them. It is unclear which proper function both determines and defines their being (225,17f.).

The functions or powers that both determine and define natural bodies can be either passive or active. In other words, the organisation something has when it is what it is might confer distinctive capacities either to do something, or to be affected in certain ways. These powers, either passive or active, enter into definitions, and therefore determine the particular definition something has, and that it is determined in the particular way that it is (for it needs to have a particular organisation in order to have the given power, whether active or passive). Thus one sense in which something is *horizetai* is the sense in which it has a proper definition (or is according to this proper definition). The other is the sense in which it has a proper form. Of course, the definition of something will mention its form, which will be a power or powers.

It is due perhaps to these dual meanings of *horizein* that Alexander says that the substance of homogeneous bodies is defined/determined according to form *and* definition. It makes sense to think of the substance of homeomers as determined according to form and defined according to definition (226,1f.). Alexander brings this discussion full circle by going on to claim that:

> Aristotle says what the definition of the homeomerous parts of animals is, and by what they are given form. He says that they are defined/determined by heat and coldness and by the motion from these. For each of these [animal parts] acquires both its solidification and organisation (= composition) by the hot and the cold (226,3-6).

Alexander continues by describing the motions of the hot and cold, and listing the passive powers discussed in *Meteor.* 4.8. The view he advances is that the hot and cold, through their movements (their powers) organise homeomers. This organising is a solidifying, and the resultant more solid thing has new (passive) capacities, such as tractility, fragility, and the rest, which are now characteristic of it. This organising through solidification, and the emergence of new capacities, is determining, and the resultant determined thing exists according to its definition.

6. Substantial change, alteration and ontology – the status of the contraries

If I am right that for Aristotle the elements are composed out of the contraries, and so the contraries are the matter of the elements, then a solution is forthcoming to some vexing problems concerning alteration and substantial change. I will sketch the problems, and then discuss the solutions that the above discussion suggests. I will conclude with some exploratory remarks concerning what Aristotle takes the contraries to be.

The first problem I wish to discuss is a familiar one. How can Aristotle account for elemental transformations given his account of the differences between alterations and substantial changes? I contend that by positing the contraries as the matter of the elements, Aristotle can account for elemental transformations in the same way that he accounts for all substantial changes. But first, let me fill in some background.

Aristotle is concerned to develop a general theory of change immune to a dilemma Eleatic in origin. He presents both the dilemma, and a sketch of his solution, in *Phys.* 1. The dilemma is as follows:

> So they [the Eleatics] say that none of the things that are either comes to be, or passes out of existence, because what comes to be must do so either from (*ex*) what is or from (*ek*) what is not, both of which are impossible. For what is cannot come to be (because it is already), and from what is not nothing could have come to be (because something must be underlying) ... (*Phys.* 1.8, 191a27-31).

This dilemma is Eleatic, with the Aristotelian addition of 'because something must be underlying'. The first horn of the dilemma denies the possibility of creation from what already is. The second horn denies creation *ex nihilo*. The strength of the dilemma depends on the assumption that these two options are mutually exclusive, and exhaustive. The generality of this puzzle is important. It is a dilemma concerning coming to be in general, covering both what Aristotle calls alteration[60] and what he calls substantial change. If Aristotle were to present a univocal solution to this puzzle, one which accounts for both alteration and substantial change, as in fact he does, then we would expect to find similar accounts of alteration and substantial change.

In *Phys.* 1.7 Aristotle considers the following ways of describing a certain coming to be:

(1) (a) man becomes musical
(2) the not musical becomes musical
(3) the not musical man becomes a musical man

Aristotle claims that the terms employed here can be categorised as follows:

60. That is, change in quality construed generally.

SIMPLE	COMPLEX
man	not musical man
musical	musical man
not musical[61]	

Aristotle makes use of this analysis in his response to the dilemma of becoming. His solution is in effect quite simple. In statements, here about alterations, such as (3) above, we have both something which underlies and persists, a man, and something which takes the place of what is not, the musical taking the place of the non-musical.

> When a simple thing is said to become something, in one case it survives through the process, in the other it does not. For the man remains a man and is such even when he becomes musical, whereas what is not musical or is unmusical does not survive, either simply or combined with the subject (190a9-13).

Aristotle goes on:

> These distinctions drawn, one can gather from surveying the various cases of becoming in the way we are describing them that there must always be an underlying something, namely that which becomes, and that this, though always one numerically, in form at least is not one. (By 'in form' I mean the same as 'in account'.) For to be a man is not the same as to be unmusical. One part survives, the other does not: what is not an opposite survives (for the man survives), but not-musical or unmusical does not survive, nor does the compound of the two, namely the musical man (190a14-21).

The error of Aristotle's predecessors was not to notice that

---

61. Much debate surrounds what it is that non-substantial simple terms pick out, such as 'musical', terms which are inherently ambiguous given the use of a neuter definite article along with a neuter adjective. Is it 'musicality', or perhaps 'a musical thing'? I side with those who favour the second possibility, since only then can the statements 'a man comes to be a musical thing/musicality', and 'an unmusical thing comes to be a musical thing/musicality', make straightforward sense.

terms which pick out different categorical items may still refer
to the same thing, that is a given substance – that substance
qualified in a certain way. A given substance is one in number,
though not one in account or definition.

This is simply a retention-replacement model. We have an
underlying thing remaining, sometimes called matter or substance, and a given privation giving way to a form, the unmusical giving way to musical (or an unmusical thing giving
way to a musical thing).

We have until now seen no mention of substantial change.
For nonsubstantial change this replacement model is simple.
A good old-fashioned Aristotelian primary substance remains,
and is the underlying thing (*hupokeimenon*), as Aristotle has
suggested throughout the earlier sections of the *Physics*. The
primary substance takes on a new form while shedding an old
one. There is, however, an asymmetry between Aristotle's
account of nonsubstantial change and his account of substantial change, one which he discusses. Interestingly, this asymmetry involves the use of *ek*, 'out of ', or 'from'. Aristotle makes
the following claims:

(1) We say 'becoming X from (*ek*) Y' in case of what does not
survive, i.e. becoming musical from not musical (unmusical).

(2) We do not say 'becomes X from Y' in the case of what
survives, e.g. becomes musical from (*ek*) being man.

(3) We do sometimes say 'X *ek* Y' in cases of what does survive
if it is a case of substantial change, e.g. a statue *ek* bronze, not
a bronze comes to be a statue.

(4) We say both 'X *ek* Y' and 'X becomes Y' in cases of
contraries that do not remain.[62]

This asymmetry tempts many to claim that Aristotle treats
alteration and coming to be differently. In cases of nonsubstantial change what comes to be comes to be *ek* its opposite, the
form *ek* its privation, while the underlying thing, the substance, is not involved in any '*ek* relations'. Yet given substantial changes that which remains is involved in an '*ek* relation'.

This has worried scholars. What motivates the difference?
In the case of substantial change there is no primary substance

---

62. See *Phys.* 1.7. I have much more to say concerning these relations in my work in
progress 'Aristotle on coming-to-be from what is not'.

which can remain as the underlying thing, so what remains must also be that from which the change progresses. There seems to be no other candidate. Yet this account of generation seems to leave Aristotle open to the second horn of the Eleatic dilemma, having something come *ek* that which is already. In considering substantial change he seems to have lost a third principle which would allow the elegant retention-replacement model to be used. Aristotle needs a solution to the dilemma for both nonsubstantial and substantial change. Aristotle realises this, and claims:

> But that substances too, and anything that can be said to be without qualification, come to be from some underlying thing, will appear on examination. For we find in every case something that underlies from which proceeds that which comes to be, for instance, animals and plants from seed (190b1-4).

Aristotle seems to want to retain the retention-replacement model for cases of substantial change. If bronze (or seed) is to be the underlying thing, the problem seems to be that of specifying the privation and form. *Phys.* 190b11-15 attempts to account both for substantial change and nonsubstantial change with the retention-replacement model:

> … whatever comes to be is always complex. There is, on the one hand, something which comes to be, and again something which becomes that – the latter in two senses, either the underlying thing or the opposed thing. By that which is opposed I mean the unmusical, by the underlying, the man, and shapeless and formless and orderless are opposed and the bronze and the stone and the gold under-lie.

Here Aristotle considers substantial and nonsubstantial change together. He suggests, I take it, that in the case of substantial change the matter is the *hupokeimenon* – the bronze, while the privation is something like formlessness or shapelessness. Aristotle explicitly makes such a claim in the following important passage:

But as for things whose privation is obscure and nameless, e.g. in bronze the privation of a particular shape or in bricks and timber the privation of arrangement as a house, the thing is thought to be produced from (*ek*) these materials, and in the former case the healthy man is produced from (*ek*) an invalid. And so, as there also a thing is not said to be that from which it comes, here the statue is not said to be wood but is said by a verbal change to be not wood but wooden, not bronze but of bronze, not stone but of stone, and the house is said to be not bricks but of bricks ... (*Metaph*. 1033a11-20).

This is a plea to consider the coming to be of a bronze statue in the same light as the coming to be of a musical man. Language, it is true, does not normally include terms for privations in the case of substantial changes, since these privations are 'obscure and nameless'; they do nonetheless act like the privations in cases of nonsubstantial change.

Aristotle is here warning us against a use of *ek* in cases of substantial change. We should not assume that since in such cases that which the *tertium quid* is *ek*, out of, remains, what remains (wood, bronze) is a substance. Indeed, in such cases we seem to have a relation of *composition* differing from the case of nonsubstantial change where we have a relation of inherence and dependence. If this is correct it is an important claim: in cases of substantial change the *tertium quid* is composed of that which it is out of. This is the use of *ek* familiar to us from *Meteor*. 4, and it points rather clearly to a solution to the problem of what plays the role of the persistent substratum in the case of substantial change in general, and elemental transformations in particular. For, as we have seen, the elements are out of the contraries in a constitutive sense. It is the contraries which persist in elemental transformations, and are the matter of the elements.

It is not just happenstance that Aristotle requires a persisting material substratum in cases of substantial change, and at the same time conceives of the elements as composed out of the contraries as matter. These theories are complementary. That elemental transformations are instances of substantial change strongly suggests, if not necessitates, that such changes are analysed as involving a persisting material substrate. Many have looked to prime matter to play this role. Whether or not Aristotle countenanced prime matter, he does not *need* to rely on prime matter to account for elemental transformations, if the account so far given is correct. I will now briefly discuss Aristotle's theory of elemental transformations, demonstrating that it is in harmony with the view that the elements are composed out of the contraries as matter, and that it is the contraries that persist as substrate in cases of elemental transformations.

I claim that only by allowing the contraries to remain and underlie various stages of elemental transformations can one make sense of the actual mechanics of these transformations, which Aristotle discusses in *GC* 2. A model which has some of the contraries remaining, while others are replaced, is not only consistent with the texts, but positively suggested by them.

GC 2 seems to give a theoretical underpinning to process-types that do often take place, for example, the transformation of bronze to zinc. I say 'theoretical' because it is important to recall that the elemental transformations which *GC* 2 discusses *may never actually take place*. The elements in their purified state do not seem ever to exist in the Aristotelian universe (at least not in the sublunary world). True, water may become steam, but this is not a transformation of elemental water, just as what we walk upon is not elemental earth. As mentioned before, Aristotle devotes a whole chapter of *GC* (2.8) to argue for the claim that every element is present in each mixed body, and that all bodies are in fact mixed bodies.[63]

Consider the following elemental diagram:

Dry	Earth	Cold
Fire		Water
Hot	Air	Moist

---

63. This helps to explain the importance of mixture in Aristotle's natural philosophy. If all transformations are in fact mixed stuffs becoming other mixed stuffs, a coherent theory of mixture is essential. A longer account of what follows concerning elemental transformations can be found in my doctoral thesis, cited above.

Here the elements are placed between the contraries characteristic of them. Opposite corners are opposite contraries. Opposite sides are composed of pairs of the most distant elements, for 'more than one thing has to change' (331b7) in order for any one side to transform into its opposite. Given the symmetry of the diagram, the following transformations can be seen as basic:

(a) any side to any adjacent side (e.g. Fire to Air)

(b) any side to its opposite side (e.g. Fire to Water)

(c) any two adjacent sides to the remaining two adjacent sides (e.g. Fire and Air to Water and Earth)

(d) a pair of opposite sides to the other pair of opposite sides (e.g. Fire and Water to Earth and Air)

(e) a pair of opposite sides to one of the remaining two sides (e.g. Fire and Water to Air)

(f) a pair of adjacent sides to one of the remaining sides (e.g. Fire and Earth to Air)

Aristotle explicitly or implicitly considers each of these transformations. He consider types (a), (b), (c) and (e) as basic, rules out the possibility of transformation of type (d), and does not consider type (f) to be basic.

Aristotle claims that some transformations are fast and easy, while others are longer and more difficult (see 331a24-7, 331b7-8). It becomes clear that opposite (b) and adjacent-adjacent (c) transformations take longer *because* more things have to change. This claim is important. Aristotle tells us that the change is quick when both termini of the change share a contrary. For instance, the change from water to air is quick, since water and air share the moist.[64] Differing from such changes are those like type (b) or (c), which are said to take longer since 'there is a change of more [contraries]' (331b6-7). The change from water to fire is an example, for both the cold and the moist of the water must perish, and that both must perish is said to be the reason why such transformations take longer.

A model which has it that prime matter is what underlies such transformations cannot explain why transformations of type (a) and (e) are quick, while those of type (b) and (c) are

64. See 331a26-b4 for a discussion of these transformations.

slower. If it is prime matter which persists as the substrate through elemental transformations, why cannot two contraries be replaced simultaneously? However, if it is these very contraries which are the material substrate for elemental transformations, it becomes clear that some one of them must persist through each stage of a transformation. When fire becomes water, either first the hot persists, and then the moist, or first the dry persists, and then the cold. If the elements are composed out of the contraries as matter, then the temporal properties of elemental transformations are easily explained. If the material substrate of the elements is some sort of prime matter, elemental transformations remain mysterious.[65]

The second problem which the composite nature of the elements can help solve is Aristotle's claim that in mixtures elements exist only potentially. This sense of existence in potentiality has gone understudied. It does not fit into the typology of potentialities found in *DA*. Philoponus realised this, claiming that the status of an element in a compound is not like a child who will grow up to be a general, nor is it like a sleeping general, but more like a drunk general unsuccessfully attempting to lead troops in battle.

Amusing as this metaphor is, it is not particularly illuminating. If fire just *is* the hot and the dry, and both of these are present in all compounds, why is fire not present actually in all compounds? Indeed, if this is true for all four of the basic chemical contraries, why is everything not just air *and* water *and* earth *and* fire actually present, perhaps somehow mixed? Apart from any metaphysical concerns such a view might raise, a physical problem presents itself. Aristotle did not solve this problem until the writing of the *GC*, although he seems aware of it in an earlier work. The problem derives from the Aristotelian doctrine that the elements, but not the contraries that compose them, have a disposition, or power, to move towards

65. Gill agrees that the contraries play the role of the persistent substrate through elemental transformations. Yet she thinks that the only sense in which the contraries are material is in so far as they persist through such changes. As I have argued, I think that the contraries persist through elemental transformations *because* they are matter for the elements, not vice versa. This is no mere 'analysis' of the contraries as matter. One should ask what sort of response to the Eleatic aporia of becoming Aristotle would be making if the contraries persist through elemental transformations, but are merely *thought of* as the matter of the elements.

their natural place. For example, fire strives to rise, while earth strives to move towards the centre of the world. This doctrine yields a problem concerning the coherence of chemical compounds, a problem compounded by the fact that Aristotle thinks that everything manifest in the sublunary world is a chemical compound. It is here that we have a profound difficulty for Aristotelian natural philosophy because of an apparent inconsistency between Aristotelian physics and chemistry. The following is from Philo's work *On the Eternity of the World*, quoting from Aristotle's work *On Philosophy*:

> Of composite bodies all that are destroyed are dissolved into their components; dissolution is then nothing but return to the natural state of each thing, so that conversely composition has forced into an unnatural state parts that have come together ... Now these parts when mixed are robbed of their natural position, the upward travelling heat being forced down, the earthy and heavy substance being made light and seizing in turn the upper region, ... in place of their natural order, they have accepted disorder, and must move to the opposite places, so that they seem to be in a sense exiles; but when they are separated they turn back to their natural lot ... (*Aet.* 6,28-7,34; Ross 19b).

Here we have the motivation for what is often thought to be a mysterious claim Aristotle makes, that in mixtures the constituent elements do not exist actually, but only potentially. If they existed actually they would actually move towards their natural places, yielding the dissolution of the compound. So they exist, but potentially.

This does not tell us *what* existence in potentiality is. Why claim that constituents while in compounds have *any* privileged ontological status? This is a question still worth asking. Why do *we* think that bronze *is* copper and tin? We have, so it seems to me, at least four reasons for thinking this:

(1) The copper and tin can be extracted.

(2) The bronze has properties suitably similar to those of copper and/or tin.

(3) The bronze was made of copper and tin, there was a

process which was both the coming to be of some bronze and the mixture of some copper and tin.

(4) At the micro-level, the bronze is copper atoms and tin atoms suitably related, i.e. crudely put, juxtaposed.

Aristotle agrees with the first three of these reasons. Instead of the fourth, he wants to claim that copper and tin exist only *potentially* in the bronze. What this seems to mean is that (1), (2) and (3) are true *and* that the chemical powers of both the tin and the copper are *actually* present in the bronze. It is the hot (along with the other powers) which *was* the hot of tin and copper which is responsible for the properties which the bronze manifests. The *potential* existence of a constituent in a compound is grounded by the *actual* existence of the chemical powers which constitute the constituents.

Aristotle's move to 'mere' potential existence seems to be meant to guard against a form of mereological essentialism when applied to constituents in compounds. Although all the parts of the tin and the copper are present in bronze, Aristotle does not want to say that the tin and copper are present (actually). This is wise, it seems to me. If we say the copper is actually present, simply because the parts of the copper, say its protons, neutrons and electrons, are actually present, then a problem arises. Would we not then be forced to say that all elements of atomic weight less than copper are also present, since all their parts are also present (owing to the fact that a given element, viewed purely mereologically, is just an element of lesser atomic weight, 'and then some')? If the four elements are present in everything and everything is composed out of them, then everything is in everything, and Aristotle's theory would be reduced to that of Anaxagoras, which he finds deeply problematic.

I now turn to two related issues. First, just what are the contraries? My account of the contraries may seem to conceive of them as 'stuffs', a mortal sin according to many Aristotle scholars. Second, how can Aristotle avoid having elemental transformations, and all substantial changes for that matter, turn out to be just alterations of the persisting matter? This is thought to be a problem particularly pressing for my view which has it that the contraries persist in elemental transformation and are matter, since there is a text that seems to rule this out, *GC* 319b20f. I will consider these problems together.

What is meant by claiming that something is or is not a stuff?[66] The answer to this question is far from clear. I will argue that talk of stuffs is, in the end, uninformative, and perhaps empty, when applied to those issues in Aristotle of concern to us here.[67] Either contraries can be stuffs in a non-problematic way, or the reasons why one might think they cannot be stuffs are in the end simply a concise way of rejecting a certain interpretative line concerning Aristotle on matter and the status of the contraries.

It is easy to list things that we think of as stuffs, but far harder to characterise what they have in common. One possibility is that a stuff is something which is named by a mass-noun, something for which it is plausible to say 'how much' one has, but not 'how many' (although this might not be the correct, or only, way of characterising mass-nouns). Note how on this reading 'Eric' is not a stuff (since proper names are thought of as count-nouns), but may be composed out of stuff. In addition, it is plausible to think that stuffs must be spatially located, if not also spatially extended. Of course, things named by count nouns may also be spatially located and extended. Stuffs, on the view now under consideration, are all those spatially located/extended things which are not picked out by count nouns. These distinctions rule out abstract entities being stuffs, at least on some interpretation of what abstract entities are. Triangularity cannot be a stuff. So, the claim that the hot, the cold, the dry and the moist are stuffs (or are not stuffs, for that matter) is *not* a claim about the four universals, but about their instances. If the hot is a stuff, instances of the hot, whatever they may be, would have to be stuffs, whatever that may entail.

If spatial location (or even extension) is all that something need have in order to be a stuff (ignoring the fact that things named by count nouns are not stuffs), then the contraries seem to be stuffs, as do many quality-tropes. My skin colour is located where I am, and is extended in space (perhaps only as a surface). My hot, or the hot of me or in me, is also where I am. So it seems that the contraries pass the 'have spatial location-

66. I thank Richard Sorabji for urging me to consider this question.
67. For an opposed view see F. Miller Jr., 'Aristotle's use of Matter', *Paideia*, Special Aristotle Issue, 1978.

extension' test for being a stuff. If one does not think that this test is the same as that which asks whether one can answer the 'how much' question, it is at least equally clear that we can talk about how much of a contrary is present. Aristotle often talks about having more or less of one of the contraries. Often in *Meteor*. 4 'more of' a certain contrary may enter some body. Of course no scale of amounts of contraries is developed, yet nothing in principle rules out having such a scale or scales.

Any other characterisation of stuffs seems to me to be far more philosophically loaded. Perhaps one conceives of stuffs as things which have, or can have, properties. Stuffs would thereby be things which are, or can be, qualified in some way. Yet on this reading of what it is to be a stuff one is perilously close to characterising stuffs as things which are, or can be, Aristotelian substrata. On this reading of what it is to be a stuff, the contraries can be stuffs if they can be the subjects of changes in the appropriate sense. Alternatively, it may be shown that they *need not* play this role, in order to be thought of as matter, or the matter of the elements. In other words, if we look carefully at what Aristotle says, the issue of stuffs v. non-stuffs disappears, and is replaced by the more global question, 'Do the contraries satisfy the test (perhaps all the tests) for being matter?' If so, then they are, I suppose, stuffs, if one of the tests for being matter which the contraries do in fact satisfy is at the same time a test for being a stuff. But this would not be an *additional* claim about the contraries.

Let me now turn to the second problem. It is important to note that if all substantial changes look like alterations on the part of matter as a subject (and so the contraries as matter as subject) we have a problem for *whatever* theory of matter we accept, whatever it is that we think of as matter, or the most fundamental matter. This is, I take it, why Gill and others wish to deny that there is *any* persistent material substratum through substantial changes. They think that if there were, certain changes which Aristotle takes to be instances of substantial change would in fact be alterations. This however is a problem that Aristotle was, I think, aware of, and he offers a solution to it. As I have said, I think that the contraries are the persisting material substrate in cases of elemental transformations. Let me sketch what I take to be Aristotle's solution.

There is a passage from *GC* with which it is particularly appropriate to begin an investigation into this problem. The passage seems to present a difficulty which must be addressed by anyone who thinks that something called matter persists in substantial change, and that this matter is qualified, or has *pathê*. In general, the problem is how Aristotle can distinguish substantial change from alteration, if he wishes to have a persistent *hupokeimenon* as subject of *pathê* in cases of substantial change – one which, so it seems, sheds *pathê*, and takes on new ones – which just is the account of alteration. For prime matter advocates, the problem is trying to explain why all changes are not just alterations on the part of some persisting prime matter. The text is often translated as follows:

> If in these cases [i.e. elemental transformations] a certain *pathos* which is a contrary persists in being the same in that which comes to be and in that which perishes (e.g. when water comes to be *ek* air, if both are transparent or cold), the other [*pathos*] to which the first changes must not be a *pathos* of this [persisting *pathos*] (*GC* 1.4, 319b21-4).

In its simplest form, the problem this text raises is this: Aristotle claims that if, in a process that *seems* to be a generation, some *pathos* (or perhaps some matter, or a *pathos* as matter) persists, and what results is a *pathos* of this *pathos* (or perhaps matter) then what occurs is not after all a generation, but an alteration on the part of this persisting *pathos*. This seems to suggest that whatever remains or persists through a real generation cannot be a *hupokeimenon* for what results from this generation, or this process will not be a generation after all, but an alteration on the part of this persisting *hupokeimenon*, whether it is characterised as a *pathos* or as matter.

Texts such as this tempt some to deny that Aristotle thinks that there is a persisting *hupokeimenon* in the case of generation. They conceive of substantial changes as changes where there is no persisting *hupokeimenon*, only persisting *pathê*. My problem is then as follows: since I conceive of some *pathê* as matter (the contraries), and since matter is a *hupokeimenon*, I seem forced into conceiving of elemental transformations (and, perhaps, all processes thought of as generations and corruptions) as involving a persisting *hupokeimenon* which has its *pathê* replaced, and so, according to *this text*, forced into having generations collapse into mere alterations.

This argument presupposes a certain interpretation of the final line in the passage quoted above: '... when water comes to be from air, if both are transparent or cold, *the other* [*pathos*] to which it changes must not be a *pathos* of this [persisting *pathos*], otherwise the change will be an alteration.' The issue is the referent of the italicised term. The argument at hand requires the 'other' to be the resulting new *pathos*. Such an interpretation is made by Williams, and Joachim. The other option is that 'the other' picks out the new substance (with Gill and Miller), the water, not, say, the cold. The argument at hand requires the *'pathos'* interpretation, since on *my* theory the cold is a *pathos* of the persisting fluid, if the fluid is a *hupokeimenon* as subject. However, it is clear that the resulting *water* is not a *pathos* of the fluid, although the cold might be. In other words, this passage is potentially problematic for me if 'the other item' is the new *pathos*, for only then does it describe a situation which my theory endorses, only to reject it.

Let me now try to demonstrate why I think that the term under discussion must refer to the resultant *substance*.[68] My argument hinges on what I take to be the proper interpretation of the passage following that under consideration, 319b25-31:

> Suppose, for example, that the musical man passed away and an unmusical man came to be, and that the man remains the same. Now if musicalness and unmusicalness had not been in themselves *pathê* of the man, these changes would have been a generation of unmusicalness and a corruption of musicalness. Therefore they are *pathê* of the man, although there is the generation and corruption of a musical man and an unmusical man. Such cases are, then, alterations.

68. I am here agreeing with Miller's translation, ibid., although he wishes to draw a different conclusion. He thinks that this passage rules out anything but prime matter being the continuant through a substantial change, including a persisting *pathos*. We differ in so far as I emphasise the '... must not be a *pathos* of this [persisting *pathos*]' clause to escape the conclusion Miller wishes to draw.

This passage is obscure, and perhaps corrupt, but I think sense can be made of it. First, notice what the passage is intending to accomplish. The earlier potentially problematic passage sets out an example of how what we take to be a generation *would seem to be* an alteration if it were falsely analysed. What follows here is an example of how a false analysis of what we take to be an alteration would make that alteration *look like* a generation. These two passages are strictly parallel. The question is whether each one gives the correct account of the process which it is *not in fact* describing. That is, is the account of alteration in the elemental transformation case correct, and is the account of generation and corruption in the musical man case correct?

The musical man case, for all its obscurities, seems to claim the following: if musicality and unmusicality were not *pathê of* the persisting man, but were somehow free-floating, we would have the generation of one and the destruction of the other. Since they are in fact *pathê of* the persisting man, we have an alteration. If they were not *pathê* of the man, if musicality simply perished, and unmusicality simply came to be, then we would have a coming to be (of unmusicality), and a perishing (of musicality), instead of what we in fact do have, the coming to be of a *musical man*, and the perishing of an *unmusical man*, which is just to say that we have an alteration of a persisting substance, the man.

This interpretation of the argument at least makes clear what other interpretations do not, that the alternative being described is one where the *pathê* are free-floating, and are not 'of' the man, not 'of' anything. This is important. If *pathê* were free-floating and not of any persisting substratum (here a man) such seeming alterations would be generations. But should we therefore understand the account presented here to be the correct account of legitimate generations? No. Aristotle's point is not that in real generations and corruptions free-floating *pathê* are created and destroyed, and that in the musical man case the only thing that prevents it from being a generation is that the *pathê* are here 'fixed' to a substrate. Surely his point is that the *false assumption* of free-floating *pathê* yields generations and corruptions where there are really only alterations. If *pathê* are not of their substrata, then all apparent alterations will in fact be cases of generation and corruption on the part of *pathê* in no way fixed or dependent on their sub-strata (*pathê* perhaps only being *where* the substrate is). This example has *pathê* acting like substances, being separate, and so coming to be and passing away.

This has ramifications for the proper interpretation of the elemental transformation case. It strongly suggests, in order to keep the two examples parallel, that the interpretation of the elemental case as an alteration *is equally a false one*, i.e. is not in fact a true account of alterations. In other words, this example claims, 'if alterations were like this (*but they are not*) elemental transformations would be alterations'. But the best interpretation of the elemental case is that the false theory of alteration is that according to which *if water* were a *pathos* of the persisting transparent, or cold, then we would have an alteration, not a transformation. It would be analogous to the musical man case, for here we would have a substance (the water) falsely treated as a *pathos* (where in the musical man case *pathê* are falsely treated as substances). I do not, there-fore, think that this passage claims that if *the cold* is 'of' the persisting fluid, the change would be an alteration, since I think that this is *the right* account of what happens in such alterations. And, more generally, on any theory which has it that there is a persisting substrate through transformations, where *pathê* are replaced, a problem would arise if we read this passage as concerning *pathê* throughout.

The false claim here is not that generations do not involve persisting substrata with replacement of *pathê*. Instead it hinges on an incorrect way of analysing the relationship of *pathê* to substances. There is, in cases of generation, a persist-ing material substrate with replacement of *pathê*, but the resulting substance is not a *pathos* of the persisting substrate. This suggests that Aristotle's real answer to how it is that generations are not alterations relies on certain features of the relationship between *pathê* and substances, and it is here that I think the solution lies.

In particular, I think that the distinction between altera-tions and substantial changes is based on *what* the *hupo-keimenon* is in any given change. If it is a substance, that which is a *tode ti*, and separable, then we have an alteration; if it is (only) matter or, worse yet, a 'mere' *pathê*, we have a generation and a corruption. It is for this reason that Aristotle goes to great lengths to deny both that matter is substance (although of course the question bothers him) and that it is separable, or a *tode ti* (related claims). These claims are meant, among other things, to prevent matter from having such a 'robust' status that the distinction between alteration and generation collapses.

As we have seen, the danger that matter may in the end be substance threatens the distinction between alteration and generation. This is a real danger because substance (or at least some plausible candidates for what it is to be substance) seems to be defined precisely in terms of being a *hupokeimenon*. If this is what matter is, then matter is substance, and the problem unique to my interpretation (that the contraries could not be matter, for substances do not have contraries) would be slight in comparison to another: that Aristotle's system would resemble strongly that of the Presocratics whom he ridicules at *Metaph*. 983b5f., thinkers who analyse all change as akin to a woman becoming musical. That is, they do away with generation and corruption. Aristotle needs there to be *various* ways in which something can be a *hupokeimenon*.

Let me sketch what I take to be Aristotle's solution. Aristotle seems to be aware of these dangers, since he offers (at least) two definitions of substance, which are related to two ways in which he thinks something can underlie. The two senses of substance appear first in *Metaph*. 1017b23-6, at the end of the entry on substance. Here we are told that substance is the ultimate (*eskhaton*) *hupokeimenon*, which is not predicated of anything else, or (and?) that which is a *tode ti* and separable. It seems that matter may satisfy the first standard. This test is applied in *Metaph*. 7.3. Here Aristotle begins by claiming that the *hupokeimenon* is that which is not predicated of anything else, and of which everything else is predicated. This is said to be a test of being a substance. Matter, form, and the composite are all said to be candidates for being such a *hupokeimenon*. Aristotle then claims (1029a7-10) that not enough has been said, and that 'this is unclear', because on this account *'matter becomes substance'*. What follows is the famous stripping example, and then the claim that matter is neither a *tode ti*, nor separable, each of which is said to be a mark of substance.

Now one might think that the conclusion to draw from all this is that matter is not therefore a *hupokeimenon* after all, for then it would be substance, and so a *tode ti* and separable, and these it clearly is not. I do not think that this is the option Aristotle takes. Instead he claims that there are *two* ways in which something can underlie, or be a *hupokeimenon*. In one way the *hupokeimeon* is a substance, when what underlies is *also* separable and a this. Matter is explicitly said to underlie, but *not* in the manner which entails that it must be substance. A substance underlies (in a particular way) and is both separable and a *tode ti*. Matter underlies (in another way) and is not separable or a *tode ti*, and so is not substance. This Aristotle makes clear at 1038b1f. Here he claims that there are two ways in which something can be a *hupokeimenon*. A *hupokeimenon* might be a *tode ti*, like an animal which underlies its attributes, or a *hupokeimenon* might be matter, which underlies the actuality (*entelekheiai*). These are two distinct ways of underlying, and two different things satisfy each test. The first is satisfied by *hupokeimena* which are substances, the second by *hupokeimena* which are matter. The sense in which matter is a *hupokeimenon* is not the sense in which it is thereby substance.

This view is echoed at 1049a19-b1. Aristotle explicitly claims here that matter (especially first matter) is not a *tode ti*, although it is a *hupokeimenon*. A substance is a *tode ti*, and is not a 'mere' *hupokeimenon* as matter is. Substances are considered to be the *hupokeimena* of *pathê*, yet when the predicate is a form or a this the ultimate subject is matter and material substance (meaning the matter of a substance). Here too matter's role as underlying is distinguished from that of substances. When we are tempted to say that the transparent, or the cold, becomes water (that is, if we try to describe the substantial change as an alteration), we should realise that the subject is only matter, and not a substance, and so the change is not an alteration in a persisting substance.

In effect Aristotle is relying on category theory to distinguish alterations from substantial changes. It is not that the mechanics of alterations are very different from those of substantial changes, but *what* changes in each case differs. The distinction is based on the differences between substances, *pathê* and matter, and so the differences between changes where each of these may underlie. A passage from the *GC* brings out this point nicely (317a17-27):

> In the substrate there is that corresponding to the definition (*logos*), and that corresponding to the matter. Therefore when the change is in these, there will be coming to be, or destruction, But whenever the change is in the affections (*pathesi*), and is accidental, it will be an alteration.

Only things that are themselves hylomorphic can either undergo substantial change or alteration. When water becomes air it is not an alteration on the part of the moist, because the hot and the cold are not affections *of* the moist. The moist *is* an affection; it does not *have* affections, even though it is the matter for the elements. Similarly, the moist cannot itself be either terminus of a substantial change, since it is not hylomorphic. It cannot lose *its* form and take on another, nor have *its* matter replaced. It is only hylomorphic *hupokeimena* that can alter or be created or destroyed; (merely) material *hupokeimena* cannot undergo change.

In fact, a passage from *Phys.* 1.7 examined above also supports this claim, at least for the case of alteration. At 190a9f. Aristotle argues that in cases of alteration the *hupokeimenon* must be plural in form, must be qualified in more than one way (a14-16). So, if something does persist through an alteration (or a substantial change, for that matter), yet is not plural in form, *it* does not alter. If an unmusical woman becomes musical, yet remains tanned throughout, this change cannot be viewed as an alteration on the part of the persistent tan, since the tan is not itself plural in form. This also explains why elemental transformations are not alterations on the part of the persisting contraries, since these contraries are not plural in form.

This way of distinguishing substantial changes from alterations, based on the differences between substances, *pathê* and matter, is also suggested by other passages in *Phys.* 1.7. At 190a39-b3 Aristotle tells us that in cases other than substantial change there must be something underlying the change. This substrate is said to be substance, 'since everything is predicated of it, yet it is not predicated of anything else'. But Aristotle immediately goes on to claim 'that substance too, and whatever can be simply, comes to be *ek* a *hupokeimenon*'. This establishes that Aristotle wants to have a univocal account of alteration and substantial change (a claim already discussed above) – both involve a persisting *hupokeimenon*. Yet it also implies that, in the case of substantial change, that *which is the hupokeimenon will be predicated of the substance*, not vice versa, since what persists is not substance and everything is predicated of substance. So, although the same account is to hold for both alteration and generation, they are distinct because of the ontological differences involving inherence and predication which hold between substances on the one hand, and attributes and matter on the other.

This emphasis on the intrinsic difference between substance and non-substance is brought to bear on the issue at hand in a famous passage later in the same chapter (191a7-12). We read:

> The underlying nature can be known by analogy. For as the bronze is to the statue, the wood to the bed, or the matter and formless before it receives form to whatever has form, so is this underlying nature to substance, that is the *tode ti* and the *to on*.

Here substance is said to be *tode ti* and *to on*, but the *hupokeimenon*, which is explicitly said to be matter, is not. It is clear that this passage is concerned only with substantial changes, and their *hupokeimena*. This theory is worked out further in the *Metaph.*, in passages such as those we considered above.

A well-argued case has been made by Scaltsas that, contrary to the view presented above, *Phys.* 1.7-8 positively rules out the possibility that the elements are composed of the contraries.[69] I will list briefly his five points. My responses to these will be a summary of much that I have argued.

Scaltsas argues as follows:

(1) Contraries (or opposites) cannot change into each other (or act on each other); what changes must be a substratum.

---

69. T. Scaltsas, 'Substratum, Subject and Substance', *Ancient Philosophy* 5, 1985, 215-40.

(2) If the elements are composed of the contraries, the contraries are principles (for the elements). Yet, the contraries are said of the elements, and so the elements seem to be principles for the contraries, and so there would be a principle of the principle.

(3) The contraries cannot be substances, since substance has no opposite. Yet if they are not substances, then substances are composed out of non-substances, which is impossible.

(4) Nonsubstance cannot be prior to substance.

(5) All change must be described with respect to a substratum and a form, yet viewing the contraries as constituents of the elements rules this out.

These worries can, I think, be addressed. First, one should note that *Phys.* 1.6 is highly polemical, and is crafted around an objection to an Empedoclean account of change. Aristotle states at 189a22 that it is difficult to see how rarity could act on density, or Love on Strife (both examples taken from Empedoclean physics). Some third thing is necessary for them to act on. We know that this objection should not be generalised, since Aristotle *does* think, in a sense, that opposites act on each other. In fact, *Meteor.* 4 is full of examples, usually of the hot and cold acting on each other. I will have more to say below about what happens to the contraries when they act on each other. Here Aristotle's objection is directed against those who think that these contraries are free-floating, and not the contraries of some third thing. Aristotle is claiming that strife cannot act on love; but the strife *of* something can act on the love *of* something else, or more to the point, the heat of something can act on the cold of something else.

As for the claim that contraries cannot change into each other, but must belong to a persisting substratum, I, in a sense, agree. However, I think that the persisting substratum can be a contrary, just not the contrary that is being replaced, or being added. Concerning the second objection, Aristotle does think both that contraries are principles, and that that which underlies is a principle. The principles for Aristotle are three. This is the conclusion of *Phys.* 1.7. I will shortly say more about the status of the contraries as principles given the account of principles found in *Phys.* 1.

Concerning the third objection, as I have suggested, Aristotle thinks that once one has the right theory of the relation of the contraries to that which they qualify, they can be that which substances are out of (*ek*), just as a form is out of a privation. It is Aristotle's analysis of *ek*, sketched above, which solves this problem. Concerning the fourth objection and the priority of the contraries, only a theory which holds that there is a temporal act of the construction of a substance out of free-floating contraries would make the contraries prior to substances in a problematic way. Yet, as we have seen, the contraries as matter do not pre-exist the elements in this way, nor do the elements pre-exist the homeomers which are composed out of them. Concerning objection 5, I agree that Aristotle wants to analyse all change with respect to a substrate and form. I do not, however, think that having the elements composed out of the contraries threatens this principle.

Taken as a group, these objections call into question the status of the contraries as Aristotelian principles. Here I will explore only one aspect of Aristotle's theory of principles which I think bears directly on the above account of the contraries, and displays a further link between *Phys.* and *Meteor.* 4.

Phys. 1.5 opens with Aristotle agreeing with his predecessors in making the contraries principles. He thinks that the principles must be contraries for two reasons (88a27-30):

(1) Principles must not be *ek* one another or *ek* anything else.

(2) Everything is *ek* them.

Aristotle claims that the 'first' contraries satisfy these conditions. Since they are first, they are not *ek* anything else, and since they are contraries, they are not *ek* each other. Condition 1 rules out any generation or destruction at the level of the first contraries, while condition 2 requires that there be a sense of *ek* such that all things are *ek* these first contraries.

Aristotle's commitment to these features of principles is forgotten by many, or so it seems. The elements clearly do come from each other (*GC* 2 discusses such elemental generation), and so do not seem to pass this test for being principles. Yet the contraries, hot, cold, dry, moist, in so far as they are contraries, do satisfy these criteria. In the *Phys.* these first contraries are characterised quite generally as form and privation. *Meteor.* 4 and *GC* give concrete examples of how the project of *Phys.* 1 works. The contraries which play the role of principles for the many changes that these texts discuss are the four contraries, which are, of course, described in terms of form and privation. This fact, more perhaps than any other, unifies *Phys.* 1, *GC* and *Meteor.* 4. Many see parts of the biological corpus as the practical instantiation of theories worked out most abstractly in the *Metaphysics GC* and *Meteor.* 4 bear this same relation to *Phys.* 1.

If the view being presented is correct, the four contraries, in their role as principles, cannot be created or destroyed. We have already looked at the *Meteor.* 4 account of what the contraries *do* when they act, but what about what *happens to them* when they act on each other? Does *Meteor.* 4 support the claim that the contraries as principles are neither created nor destroyed?

The answer to this question, perhaps surprisingly, is yes. In all of *Meteor.* 4, the four contraries are never unambiguously said to be created or destroyed, nor is any one said to come from another. Aristotle has a rich vocabulary for describing true generations and destructions. None of these terms are employed to describe what happens to the contraries during the descriptions of the many *pathê* discussed. They are said to be absorbed (380b19-20), driven out (*ekkrouentai*, 381a16, *sunexatmizontos*, 382b21), trapped inside (381b2-4), absent or present (*apousiai ê parousiai*) (382b1), concentrated and compressed (*sunagein ê antiperiistanai*) (382b10), drawn out (383a17). They are said to leave (382b27, *aphairethentos*), evaporate and condense (383a12), return (383a28f.), lose their strength (*marainomenou*) (383b31), vaporise (384b9f.), and escape (388b15). None of these seem to be equivalent to destruction or generation.

All this supports the claim that the contraries, as principles, are neither created nor destroyed. It is also supported by the fact that *Meteor.* 4.12 claims that all the *pathê* of homeomers are due to the four contraries and their motions.

This account might strike one as 'too mechanistic' for Aristotle, since it suggests that many of the *pathê* of homeomers are due to the entering, leaving, and other motions of contraries. I will end by discussing briefly an interesting question this raises: just how Democritean is Aristotle? I think he may be quite Democritean, perhaps more so than even he realises. For instance, consider the *pathê* discussed in *Meteor.* 4.8. Melting

and softening are related to the pore structure of bodies, and
to whether their parts are dispersed by entering liquids. Flexi-
bility is a change of shape, a spatial rearrangement of the
whole. Breaking and fragmentation are spatial changes, and
said to be due to pore arrangements. Impression is an inden-
tation, appropriate to squeezable things the parts of which
somehow contract into themselves (because their pores are
empty). Ductility, malleability, fissility, and the cutable are all
accounted for via spatial changes of the parts of a body. The
viscous is that which has its parts arranged like links of a chain.
The friable has parts that lack this structure. Compressible
things can be squeezed, and so contain pores. Even combustible
things are those that have long pores into which fire can
penetrate.

The case with the more 'global' *pathê*, drying, thickening and
solidifying, is a bit more difficult. 'Increasing in density' is said
to be due to the dry constituents of a body packing together
more closely (see 383a12f.), and this increase in 'density' is
closely related to what happens in the solidification of non-
watery fluids. Since many *pathê* result from the solidifying of
a body, if solidifying is a topological change, these other *pathê*
would at least require such a change for their existence.

Aristotle's Democriteanism arises, I think, from another
angle. If the contraries are material components, and drying,
solidifying, and the like, are due to various contraries' leaving
and/or entering (as *Meteor.* 4 has it), then these *pathê* (solidity,
etc.) are due to the addition and/or subtraction of matter. This
too is Democritean. Again, the question would be whether a
change in the matter of an object in a certain way is what *it is*
to be solid, or just is required for solidity to arise. Of course the
distinction between the two is not always clear. Democritus is
'an eliminativist reductionist' (by custom sweet … in reality,
atoms and void[70]), yet this is not the only position one might
take concerning the relationship between supervenient properties and that on which they supervene. Aristotle might
believe that solidity requires (or perhaps itself causes) rearrangement of the parts of that which is solidified, but not that
solidity can be reduced to this arrangement. Yet even this much

70. DK B9 = Sextus *AM* 7, 135.

has Aristotle more Democritean than many would allow.

I think that Aristotle's position with respect to atomist explanations of generations is best summed up by a passage at the end of *Meteor*. 4. Here Aristotle shows his sympathy with mechanistic explanations of many *pathê*, yet at the same time clearly states what he sees as the limitations of such theories.[71] With this passage I shall end this introduction, hoping that I leave you curious concerning, if not also convinced of, the importance of *Meteor*. 4.

> Since solidifying is due to the hot and the cold, heat and cold and the motions arising from these are sufficient for producing the parts such as these. I mean the homeomerous parts such as flesh, bone, hair, sinew and others such as these. For all of these are differentiated by the differentiae we have already discussed, tensility (*tasei*),[72] tractility, fragility (*thrausei*),[73] hardness, softness, and the rest. For these are produced by the hot and cold, and the mixture of their motions. But no one would think this of the anomeomers (for example, a head, hand, or foot) which are composed out of these homeomers. For although coldness and heat and their motions cause the production of bronze or silver, they will not account for the production of a saw, a cup, or a box. In these cases craft (*tekhnê*) is the cause, in others nature, or some other cause (390b2-14).

Acknowledgements

Many have helped in seeing this project to fruition. Very early drafts of the translation were read by M.L. Gill, J. Lennox, and V. Caston, whose comments pointed me in the right direction. R. Sharples looked at assorted stages of the manuscript, and his unparalleled knowledge of Alexander both rescued me from many a blunder, and suggested many an improvement. J. Ellis discussed some vexing passages with me, as did my colleague

---

71. Of course sympathy with mechanistic explanations is distinct from sympathy with full-fledged atomism. Aristotle is in no way sympathetic with the Atomist's postulation of the void, or of indivisible atoms, the two cornerstones of atomism.

72. This is a general term meant to include a number of the passive powers discussed in *Meteor*. 4.8-9.

73. This term seems to incorporate both things that break and those that shatter.

S. Menn. D. Furley commented upon a section of the final draft, much to its improvement. D. Russell sent me a set of extremely learned and useful comments. Many of his suggestions I have incorporated into my translation, the more substantial of these credited in footnotes. H. Blyth's expertise in ancient technology came to my aid in attempting to make sense of the many technical terms referring to the properties of homeomers found in this text. My colleague Marguerite Deslauriers took on the task of commenting on the whole of the manuscript. There is not a page of what follows that was not greatly improved by her suggestions. Throughout the time I was working on this project Richard Sorabji was a constant source of advice, encouragement and philosophical illumination. He has been for many years, and continues to be, both trusted mentor and welcome friend.

I wish to thank Ian Crystal profusely for compiling the Greek-English Index.

Translator's Note

In 1936, Victor Carlisle Barr Coutant privately published his doctoral dissertation of the same year entitled *Alexander of Aphrodisias: Commentary on Book IV of Aristotle's Meteorologica*. This was a ground-breaking work, the first published translation into English of a text of the Greek Commentators subsequent to the completion of the editing of these works in the *CAG*. It seems only fitting that this series of translations from the *CAG* should honour Coutant's pioneering effort. I gratefully acknowledge that the following translation is indebted to Coutant's, yet it is far enough removed from his text in many ways, both small and substantial, that I must take full responsibility for the results. Where I have retained a footnote of Coutant's, I mark this fact. All other notes are mine.

Alexander of Aphrodisias
On Aristotle Meteorology 4

Translation

The Commentary of Alexander of Aphrodisias on Book 4 of the *Meteorology* of Aristotle

CHAPTER 1

The book entitled 'the fourth' of Aristotle's *Meteorology* does belong to Aristotle, but not to the treatise on meteorology, for the matters discussed in it are not proper to meteorology.[1] 5 Rather, so far as the matters discussed are concerned, it would follow the *Generation and Corruption*. Having there discussed the four tangible powers (*dunameôn*), heat, coldness, dryness, moisture; and having shown that from the coupling of these powers the elements come to be; and having said that some of the powers are active, and others passive; in this book he says 10 what each of these powers does or what it undergoes and what things come to be by means of these powers.[2]

> Since the causes of the elements[3] have been determined
> to be four ... (378b10).

Aristotle is reminding us of the matters set out in the second book of the *Generation and Corruption*.[4] There he has shown that there are four causes according to which the first bodies,[5] 15 the elements, are given form (*eidopoieitai*):[6] these causes are heat, coldness, dryness and moisture. They were shown to be the first tangible contrarieties,[7] and by the first tangible contrarieties the elements are given form. (By being tangible they are actually existent [*en hupostasei*] and thus differ from a mathematical body.) For this reason the elements number four, because the things by which they are given form are also four. 20 For there are six possible combinations of pairs of these four [form-giving things], of which, since two are non-existent be- 180,1 cause they consist of opposites,[8] four remain by which the elements are given form. Fire [is given form] by heat and

dryness, air by heat and moisture, water by coldness and moisture, earth by coldness and dryness.[9] It is not possible for a thing to be given form by opposites so that it is at once hot
5 and cold, or dry and moist.[10]

Of the four, he says that two are active, heat and coldness, and two are passive, dryness and moisture.[11] Aristotle said this in the *Generation and Corruption* also.[12] He offers evidence that two of the foregoing are active and two passive, first through induction (*epagogês*), and second from the defining account (*tou logou tou horistikou*).

10 [He shows] through induction that 'heat and coldness are observed in all cases to determine, unite and change things'.[13] Now he says that 'to determine' is to concentrate and thicken, or to solidify, so that what was previously without its own shape because of moisture and always took its shape from its surroundings (*hupo tou periekhontou*) assumes a shape of its own
15 and has its own limit. Moist things, such as air and water, which, because of their moisture, do not have a proper shape, are called undetermined;[14] determined things have a proper shape and do not conform to the surroundings.[15]

At any rate, water, which is undetermined, is determined when solidified by the cold, and assumes a proper shape. Moist things thickened by the hot are determined, as honey by fire,
20 and in general whatever becomes thick by boiling assumes a proper shape so that it does not conform to its surroundings.

It is clear, moreover, that the hot and the cold also unite, that is make one (*sumphuousi kai henousi*),[16] determined things. For fire combines and unites homogeneous[17] things with one another, removing and separating all foreign matter from them. This is the 'peculiarity' (*idion*)[18] of fire. For example, it
25 removes the foreign [matter], whether this is silver or copper, from gold, and unites its parts with one another. But the cold unites not only homogeneous but also heterogeneous things, for by solidifying it unifies all things that are close together and juxtaposed.

That the hot and cold can bring about change is apparent, for they moisten, dry out, harden and dissolve [things]. Some
30 things are made harder by hot, others by cold, and moreover some are made moist by each of them. But some are also softened by fire, obviously because of the heat from it.

Just as the hot and the cold are observed to determine, unite and change things, so the moist and the dry appear to undergo these processes. For in undergoing these [processes] by means of the hot and the cold, the moist and the dry are determined. The moist is determined when it is thickened and solidified, 35 the dry and hard are softened when they are loosened.[19] The 181,1 moist and the dry are seen to undergo the aforementioned affections [that is, they are determined, united and changed] from the hot and the cold both when by themselves and when something is mixed out of both of these.[20]

Having shown by induction and observation that some things are active and others passive, he says that it is clear also 5 from the accounts which we use in 'determining'[21] their [i.e. the active and passive] natures. For in defining the hot and the cold we call them 'aggregative'. We say that the hot is capable of aggregating homogeneous things but capable of separating different and foreign things; we say that the cold is capable of aggregating all things both homogeneous and foreign to each 10 other. But aggregation and separation are actions, and so action is in the account (*tôi logôi*) and the being (*têi ousiai*)[22] of these. The dry and the moist, however, we define by the words 'easy to determine' and 'hard to determine', which refer to 'undergoings' (*tou paskhein*).

Having shown that [two] of the four [powers] by which the first bodies are given form are active, namely the hot and the cold, and that [two] are passive, namely the moist and the dry, 15 Aristotle next says that 'it is necessary to understand the productions which the active powers produce, and the forms of the passive ones'.[23] That is, [it is necessary to understand] which operations (*erga*) are those which come to be by means of the active powers and what their differences in form are, and which affections the passive powers undergo and again what differences in form apply to them.

'In the first place, and generally, simple generation and natural change is the work (*ergon*) of these powers.'[24] Genera- 20 tion in general and natural change, he says, come to be by them and in accordance with them. Aristotle said 'simple generation' instead of 'generation in the proper sense, without addition', for alteration is generation of a sort. He added 'natural change' because simple generation is not the only natural change. 25

There are other changes which come to be by these powers: the growth and the ripening of fruit and other such things are changes, for example. Not only generation but also its opposite, destruction, (when it comes to be naturally) comes to be by these powers. For to be destroyed is [also] natural to every generated thing.[25]

30   Destruction which occurs naturally, i.e. which is brought about by the coming of old age (as well as by weakness and chilling), occurs in this way. But not violent destruction as when one is struck by a sword and perishes. 'These [processes] then, are inherent in plants and animals and their parts.'[26]

Aristotle speaks of the four powers according to which generation and destruction occur. For each of the things mentioned
182,1   partakes of all the powers and does not, like the elements (that is the simple bodies) contain only two of them, as has been said. He next adds [an account of] what simple and natural generation is. He says that it is a change produced by the active powers
5   (clearly heat and coldness), which change the things whose nature it is (*pephukota*) to be affected by them; these he called matter[27] (they are dryness and moisture[28]) because they have a power and predisposition to be affected, when, of course, they are in such a ratio (*logon*) that the one [pair] can act and the other be affected. It is not that any amount of hot moistens or dries any amount of dry or moist, but rather that a certain ratio and proportion is required.[29]

10   Having said that natural generation is a change in the material passive powers by the active ones, which occurs when passive and active powers are in a certain ratio to one another, a ratio in accordance with which some of the powers are able to act, and the others are able to be affected naturally by them (this is how plants and animals come to be; the passive powers
15   in[30] the matter are affected and changed by the active), generation of the thing in question occurs, he says, when the active powers master the passive and material ones. But when the active powers do not master the passive but are weakened through disproportion either because they are weaker and less in amount, or because the passive are stronger and more numerous, then there is partial simmering[31] or inconcoction.
20   Aristotle either (1) says 'simmering' of things boiling (when fire does not master them, they do simmer and do not boil, as he

will say shortly) and says 'inconcoction' of the nutrition of animals when it is not mastered by the heat in them [and made] into something useful, but is spoiled; or (2) he used both terms in the same way, making 'simmering' the equivalent of 'inconcoction'. The addition of 'partial' indicates that if such a weakness of the active power is somehow partial so that the whole   25
is not destroyed, but some of the parts assimilated to the whole [are so destroyed], and because of the change in those parts the substrate is either increased or nourished, then the weakness of the active power with respect to these parts produces inconcoction and simmering.

Having stated that the partial weakness of the agent (*tou poiountos*) produces inconcoction, he says that putrefaction   30
(*sêpsis*) is opposite and common, not to a given generation, nor to partial generation, but to simple generation. For just as all simple and natural generation [occurs] because the passive [powers] are affected by the active ones, the latter mastering, the former being mastered, so too every natural destruction occurs by putrefaction when the activities fail to master the   183,1
matter. Every natural destruction, he says, is a 'path to putrefaction', as, for example, old age in animals or withering in plants. For both are putrefaction. The natural end of every compound organised (*sunestôtôn*)[32] by nature is rottenness and putrefaction unless the destruction should be a violent one, and such destruction is not natural. For rotting flesh is not destroyed in the same way when it is burned, nor is bone, nor any   5
other part of animals. The natural destruction of them all is putrefaction.

As a sign (*sêmeion*) that the natural destruction of [every compound structured by nature] is putrefaction, Aristotle points out that at first [such things] are full of moisture [and then they become dry].[33] For they are generated in the beginning from the moist and the dry; since the dry is determined   10
by the moist and takes on a boundary and shape of its own because of its mixture (*mixei*) with this [that is, the moist]; this is brought about by the active powers, which are heat and coldness.

So much for generation; destruction occurs when the determining factor, which is the agent (this is the hot especially), is mastered by that which is determined by it, namely the dry.

15   (The dry is what is determined because of its mixture with the moist, for the dry is undetermined, not cohesive but rather prone to disintegrate unless it has some moisture). When, then, what is determined and passive masters what is determining and active because of the surroundings (because, that is, the greater amount of heat in the surroundings consumes (*analiskein*) the lesser amount in the thing which is destroyed), [then destruction occurs]. Hence things undergoing destruction are first moistened, and this is putrefaction. For putrefaction

20   is the change of the substrate into the moist and is produced by the surroundings due to a lack of proper heat, as Aristotle will show.

Having stated that everything destroyed by nature is putrefied, Aristotle says that in a special sense putrefaction is used of that which is partially destroyed when it is separated from (that is [*kai*] departs from) its proper nature, which is the

25   moving principle (*kinêtikês arkhês*) in it.[34] The hot in each thing is such [a principle]. Hence, he says all things are putrefied except fire, because the hot is the most active [principle]. That which is given form by the hot does not putrefy, because putrefaction occurs when the passive masters the active in the putrefying thing, but in the case of fire it is not possible for the hot in it to be overcome by some one of the [other powers] in it.

30   For fire is fire in virtue of this. Air, for instance, even if it is hot of its own nature, is so secondarily, and is given form more by the moist than by the hot. Furthermore, the destruction of the hot in the moist is putrefaction, but there is no moisture in fire. Although earth does not have moisture in its own nature, there is always some moisture in it, by which it is held together.

Aristotle now defines what putrefaction is and says that it

35   is the 'destruction of the proper and natural heat in any moist

184,1   thing by foreign heat';[35] and that this foreign heat, which destroys the proper and natural heat in moist things by being greater in quantity, is in the surroundings.

Consequently, since it undergoes [putrefaction] due to a deficiency of [the hot] ... (379a19).

5   Aristotle has said that all natural destruction is putrefaction, as well as what putrefaction is, and has shown through his

account that the hot in the surroundings is the cause of putrefaction. Since he laid down that there are two active [powers],
the hot and the cold, he also tells us how the cold is a cause of
putrefaction. For since putrefying things are destroyed because
of a deficiency of their proper and natural hot and since a thing   10
is cold in so far as it is deficient in the hot power, both the hot
and the cold are causes of putrefaction. Hence, putrefaction is
an affection common to both proper coldness and foreign hot.

'Because of this all putrefying things become drier.' With
these words Aristotle gives the reason why putrefying things
begin by being moist and then become dry. For the proper hot   15
while departing and dissipating takes along with it, and evaporates, the moist in it, and there is no longer anything to draw
in any other moisture. For the proper hot in each thing attracts
moisture, drawing it from without.

He also shows that it is consistent with the foregoing that
things putrefy less in cold weather than in hot. In that case,
the lesser amount of the hot in the surroundings does not   20
master the hot in each thing. But in the summer there is a
greater amount [of hot in the surroundings].

Aristotle also cites the reasons why neither the frozen nor
the hot and boiling is putrefied: that which is frozen because it
is colder than the surrounding air is hot, is not mastered by the
surrounding air in such a way that the hot in the frozen thing
is drawn out.[36] And that which putrefies and changes a thing   25
does so by mastering it and setting it into motion. But the hot
and boiling is not mastered by [the heat in the surroundings],
because it contains more heat than the surroundings. For the
greater amount of the hot, by mastering the lesser, extin-
guishes and destroys it.[37] And that which is in flux and motion
is putrefied less than that which is not moving.[38] Hence standing waters are more liable to putrefaction than flowing ones,   30
for, since the inherent heat in flowing and moving bodies is
kindled by the motion, the heat in the surrounding air becomes
relatively weak.[39] Due to the same cause a greater amount is
less putrefied than a smaller amount. For the smaller amount
is more inclined to putrefaction because it has less heat and is
more mastered by the heat in the surroundings. In the greater   185,1
amount, he says, more of the hot and the cold exists within
(*enuparkhein*); hence the greater amount is not easily putrefied

by being mastered by the heat in the surroundings, nor does it undergo from coldness what things affected [by coldness] undergo from the overpowering of coldness.

5 For this reason he says that the sea when broken up into parts quickly putrefies. At any rate, the pools cut off from it do so, but the whole sea does not. Similarly, small bodies of other kinds of water are putrefied more easily than large ones.

Aristotle gives the cause of the generation of living creatures such as worms, mosquitoes, and gnats in putrefying things. 10 [Namely,] because the heat separated off from these putrefying things is natural, it structures (*sunistatai*) the bodies separated off with it, provided they still have some moisture.

We have said what generation and destruction are (379b8).

Aristotle summarises his remarks. Generation has been said to be a change of the underlying matter [brought about] by the 15 active powers in their relation to (*kata*) the passive powers. This occurs when the active powers are in such proportion to the passive powers that they master them. Yet in the *Generation and Corruption* Aristotle characterised generation as a change from being in potentiality to being in actuality. There he gave a more general and universal account of it; here the account is more particular and physical, if indeed the inclusion 20 of matter (in which the things which come to be by nature come to be) in the definition is a mark of the physical.[40]

Destruction is the reverse. It occurs when the passive [powers] master the active [powers] by means of the surroundings, for this is 'the path to putrefaction'. Every natural destruction is putrefaction, and putrefaction occurs in general in this fashion.

CHAPTER 2

It remains to discuss the next forms, those which the
aforementioned powers produce (*ergazontai*) out of the   25
substrates that are already naturally constituted
(379b10).

Having discussed generation and destruction, it remains, he
says, to speak of the forms which follow on the aforementioned
[forms], (namely on generation and destruction), which themselves come to be by means of the powers we have discussed.
These also are the [forms] through which the previously mentioned powers produce each of those things which they produce
by acting on the already constituted natural substances   30
(*ousiais*). The bodies of animals and plants are such [structured
natural substances].

Well of these, common [i.e. generic] ones – i.e. those through
which every one of the effects in the foregoing [passage] comes
about – are concoction and inconcoction, the one being analogous to generation, the other to destruction. Concoction is due   186,1
to the hot and inconcoction is due to the cold, and each of these
[i.e. concoction and inconcoction] chiefly [comes to be] through
each of those [i.e. the hot and the cold].[41]

Then subordinate to each of these two [generic ones], concoction is subdivided into ripening, boiling and roasting (for, while
differing from each other in form, they are all concoctions),
inconcoction is subdivided into rawness, simmering, and searing (for each of these is an inconcoction). As inconcoction is the 5
opposite of concoction, so too what falls under these are [op-
posed], rawness to ripening, simmering to boiling, and searing
to roasting.

Having stated that the forms of concoction are ripening,
boiling and roasting, those of inconcoction rawness, simmering
and searing, Aristotle says that we must realise that these
names are not proper to the things according to which they [the
names] are grasped, but since there is no one common and   10
general name laid down for all similar things, the name apply-

ing [properly] to some is transferred to every similar thing, as we shall see later in our study.[42]

Aristotle next says what each of the foregoing [processes] is. First he says what concoction is. Concoction, he says, is 'a perfection by means of the natural and proper hot out of the

15 contrary passivities'.[43] In the case of animals the perfection of the nutriment by the nutritive soul, through the natural proper heat, so that nutriment in potentiality becomes [nutriment] in actuality, is a concoction. And also in the case of all other things, concoction comes about by means of the hot in those things [which are] being perfected. Aristotle said 'out of the

20 contrary passivities' instead of 'out of the underlying matter', which is changed by the hot into [having] affections opposite to those which it possesses [prior to this change]. For nutriment has two modes of being, the first prior to concoction, the last after concoction, and the two are contraries. Hence in the *On Generation [and Corruption]* Aristotle said that something which is nourished is nourished in one sense by its contrary, in another by something similar [to it].[44] For the still unconcocted

25 nutriment which we eat is the contrary of that which is being nourished, but the nutriment which exists after concoction and which is already being assimilated and is nutriment in the strict sense is similar [to that which is being nourished].[45]

For while everything being concocted comes into being and is concocted by alteration and change, the change is out of contrary into contrary. The matter which is being concocted, he said, is opposed to that which has been concocted, perfected

30 and brought into form from it. He says that the matter being concocted is passive (as every such thing is), for its perfection comes from being affected. Aristotle, in explanation of the foregoing, added 'for these (meaning the opposite passive [powers]) are the matter proper to each thing'.[46] For there is not the same substrate in all [things] being concocted,[47] nor are the same things concocted in all cases, but when concocted things

35 are concocted they are perfected and that which was in potentially comes to be in actuality.

187,1 The principle (*arkhên*) and cause of perfection is the proper heat in the thing being nourished, he says. For nature uses this as the principal tool in bringing about concoction, although the aid of external factors may help bring concoction to perfection.

For many external factors seem to help the concoction of nutriment, such as baths and exercises, but these, by making the things being concocted easy to digest and easy to change, are [but] aids to concoction, while the principle and cause [of concoction] is the heat in the thing which is being nourished.

Having said that concoction is a perfection, Aristotle says what the perfection and end of that which comes to be out of concoction is. In some cases, he says, the end of concoction is the [concocted thing's] nature. Since nature is dual, [being] on the one hand as matter, on the other as form, he adds what sort of nature the end of concoction is; it is as form and substance. For the substance of each thing and its being what it is are in virtue of (*kata*) the form of that thing. In some cases, then, he says the end of concoction is to realise the natural form itself, as in the case of fruits. For their concoction is like that of nutriment, and the end of nutrition is the natural change [of the nutriment] into the form of the thing being nourished.

Since concoction is said not only of nutriment, and since we also say that certain other things are concocted (such as must, tumours and other such things) in the case of things concocted in this fashion, he says the end of concoction is the change into a certain form (*morphên*) in which the thing coming to be is more useful and more beneficial[48] to us.[49] Describing what the underlying form (*morphê*) is, Aristotle says: 'when the moist comes to be of a certain quality or quantity by being roasted, boiled or putrefied,[50] or heated in some other way', so that it is useful, then we say that it has been concocted. This is how we say must and the structures in tumours are concocted. We say that must has been concocted when it comes to be wine, the structures in tumours when they come to be pus, and the discharges in the eyes when they [come to be] rheum. The concoction mentioned in connection with phlegm is similar.

To suffer this happens,[51] says Aristotle, to all things being concocted, [namely] that they are concocted and complete their concoction when their underlying matter and moisture have been mastered by the natural heat. He says that as long as the hot in all such moist things is in such a proportion to the moist that it is able to move and change this moist, this is its nature. For the essential (*kath' hauto*) (and not the accidental) source

5

10

15

20

25

30

of motion in each thing is the nature of that thing, as was shown
in the *Physics*.[52]

188,1   Aristotle says that concoction is constituted by the proper
hot mastering the moist. Hence, he says, the following are signs
of health, that is, things said to have been concocted: urine,
faeces, and residues in general, because the concoction of these
5   is a sign that the proper heat in the animal is mastering the
moist, for this [the moist] is the undetermined [matter]. He also
says what the concocted things come to be like. They must, he
says, be thicker and warmer than unconcocted things, for the
hot makes (*apotelei*) them hotter, thicker, more compact and
drier.

Having said what concoction is, Aristotle next discusses its
10   contrary, which is inconcoction. He says that inconcoction is an
imperfection due to a lack of proper heat. The lack of proper
heat is proper coldness; hence inconcoction is an imperfection
caused by proper coldness. Just as when having said that
concoction was a perfection he added of what it was a perfection
(namely of the opposed passivities, and these were the underlying substrata for concoction, being matter), so too having said
15   that inconcoction was an imperfection, he added of what it was
an imperfection, namely, of the opposed passivities, which are
the matter underlying the affections and the changes in each
concoction.

CHAPTER 3

Having said what concoction and inconcoction are, Aristotle
goes on to discuss their subsidiary forms. He first says what
. 20   ripening (*pepansis*) is. He says that ripening is a concoction in
kind (*genei*). In defining ripening he showed that this is so. He
defined ripening as a concoction of the nutriment within pericarpia. (He says the pericarpia are what surround the fruit
[*karpois*] in which there is a certain moisture, the concoction
of which is ripening.)[53]

'Since concoction is a perfection (*teleiôsis*) ....'[54] Taking rip-
25   ening as a concoction, and having set out in the definition of
concoction that it is a perfection, Aristotle states when ripening
is perfected: when the seeds (*spermata*) in the pericarpium

have the power (*dunêtai*) to make (*apotelein*)[55] another like it
[i.e. the pericarpium].[56] And with other things too, each is said
to be perfected (*teleion*) [in a similar sense]; for example, a
perfected person has the power to generate another like himself, as does a horse, and each of the other [natural kinds].

Such is the ripening in pericarpia, and ripening seems to be   30
its proper (*kurion*) name. But many other things are said to
have been concocted and to be ripe, coming to be [so] in a process
similar to (*homoeidôs*) the concoction in pericarpia, but not
properly having the name 'ripe'. By metaphor they are so
named, after the ripening in pericarpia. As Aristotle has said
previously, there is no general name for each of the forms
falling under concoction and inconcoction. But because one of   189,1
the subsidiary [forms] is named in this way, the rest are also
so named by similarity and metaphor.

Having stated that certain [subsidiary forms] are named
metaphorically because no proper names have been set down
for each subsidiary form of concoction and each perfection of it,   5
Aristotle added, 'with respect to that which is determined
(*horizomena*) by natural heat and coldness',[57] meaning not that
some concoction is also produced by coldness (for it has been
established that concoction comes about by heat), but that
coldness also contributes (*suntelousês*) to concoction. For the
heat determines the moist and the cold as matter by mastering   10
[them] in concoction.[58]

Aristotle showed that the other [things], to which on the
basis of similarity he metaphorically attributed the name 'ripening', come to be and ripen in a manner similar to the things
[ripened] in pericarpia, by saying: 'The ripening in tumours,
phlegm and the like is'[59] the concoction of the moist in them by
the hot inherent in (*enuparkhontos*) them, for the moist cannot   15
be determined unless it is mastered by the hot. The nature of
ripening in pericarpia was [said to be] such.

'Out of airy things (*pneumatikôn*)[60] ....'[61] Aristotle states that
ripened [things] become watery from being airy. For pneuma
is condensed (*sunistatai*) and thickened (*pakhunetai*) in the
change to water, and watery things are [structured and thick-
ened] in turn, in the change to the earthen. To the degree that
watery things are thickened and become more earthy, to that   20
degree they are said to have been concocted. In general

(*katholou*), all ripening things become thicker from being thinner.[62]

The phrase, 'nature incorporates (*agei*) some things into itself in this process and discards others',[63] might mean this: of ripening things, some are changed into a natural form, in so far as (*kath' hoson*) they are ripened, and the change into this [natural form] is the natural end of their ripening. Other [ripening things] are cast out and excreted from the things 25 which have them by nature, as, for instance, phlegm, pus, rheum, and excreta (*perittôma*).

Aristotle has described what ripening is. He next discusses rawness (*ômotêtos*), which falls under 'inconcoction' and is the contrary of ripening. He states that rawness is the contrary of ripening. For what is properly called 'rawness' is an inconcoction of the nutrition in pericarpia, which is the undetermined 30 (*aoristos*) moisture [in them], this being unorganised (*asustatos*). Hence he also says that rawness is either airy, watery, or a mixture of both. If that which should have concocted is air, [the rawness] is airy; if it is water, [the rawness] is watery. It can be out of both when that which is concocting should thicken, but remains airy and watery.

Since ripening is a perfection, it is clear that rawness will be 190,1 an imperfection (*ateleia*) coming to be because of a lack of natural hot and its disproportion (*asummetrian*) to the ripening moist.[64] For when the hot is too little to be able to master the moist, the moist remains unconcocted.

Since Aristotle has stated that it is the moist which is ripened, he now says which of the moist [things] are ripened: 5 namely all that are capable of thickening when heated. For everything which is ripened is condensed[65] (*sunistatai*). But no moist [thing] unmixed with something dry and earthy thickens, and therefore it does not ripen.[66] Necessarily, then, the ripening moist thing also contains in itself some dry admixture. Hence water alone of the moist things does not thicken when heated because it does not contain in its substance (*en têi* 10 *ousiai*) any dryness; when heated it evaporates and is destroyed, more quickly than it can thicken. Olive oil and wine thicken when heated because they do have in their substance something both dry and earthy.

After saying that no moist thing unmixed with something

dry is ripened, he speaks again of rawness and the imperfection
in it. For it was laid down that rawness comes to be because of
a 'lack of natural hot and its disproportion to the moist being
ripened'.[67] This disproportion of the hot to the ripening moist 15
comes about, he says, either because the hot is inconsiderable
(that is, less than that which is according to nature [*kata
phusin*]) or because the moist being concocted and determined
is considerable (that is, more than is according to nature). For
this reason he says that both the juices and pulps (*sustaseis*)[68]
of raw things are thin [in consistency], cold rather than hot,
and inedible and undrinkable, since [in raw things] the moist 20
has not been mastered by the hot.

Just as 'ripening' was spoken of in many ways (the name
applied properly to the ripening of fruit, but certain other
things were said to ripen as a metaphor from these cases,
because the process is similar in them, and that which comes
to be in their case is similar [*homoeides*] to that [which comes
to be] in fruits), so, says Aristotle, 'rawness' too is spoken of in 25
many ways. For rawness is attributed not only to fruits but to
the other things to which ripening also is attributed. Urine and
evacuations (*hupokhôrêseis*) from the belly [i.e. vomit][69] and
effluxes are all called raw because they have not been mastered
by the proper heat nor have they been condensed.

Moving ahead, we call pottery, milk and many other things 30
raw too. For whatever is capable of changing (*metaballein*) and
being condensed (*sunistasthai*) by heat, yet is not condensed,
nor is affected, is called raw. For we do not say that things
unable to be condensed by heat are raw. Water is not called raw
because it is not naturally thickened by hot.[70] Aristotle says
that it is clearly said to be boiled and unboiled, but not raw or
ripe.

It is clear that pottery is more raw than other things and 191,1
that it is not called [raw] in a way altogether similar to [the
way] in which these [(fruit, urine, etc.) are called raw].[71] For
things [e.g. fruits] ripened by means of their proper heat were
raw and were called raw because the proper hot in them did
not master the moist, but pottery is called raw because the hot 5
applied to it (*prosagomenon*) from without has not mastered
the moisture in it. One might ask further how milk also is said
to be especially raw and not in a manner similar to them [i.e.

fruits etc.]. Now if milk is called raw when in the animal, it is raw in the same sense as the primary cases [i.e. like fruit, etc.], for it is raw because the moisture in [the animal] has not been

10   mastered by the heat in the animal.[72] If milk boiling over a fire is called raw, it is called raw in a way similar to pottery, for it is raw because it has not been mastered by the heat from without.

Aristotle has spoken of ripening and rawness, the first of which was classed as a concoction, the second as an inconcoction, and has defined them and explained the causes because of which they occur. He next discusses boiling and its opposite

15   simmering,[73] treating boiling first.

He says that boiling is in general a 'concoction of the inherent undetermined [part] in[74] the moist by moist heat', using the phrase 'moist heat' because it is by the hot-moist that things are boiled. For things which are boiled are boiled by hot water and in hot-moist.[75] 'Of the inherent undetermined [part]'

20   means of the undetermined and watery [part] in the moist which is present in boiling things. For the moist in these things, being undetermined, after being concocted and determined by the aforementioned heat, comes to be boiled from the boiling [it underwent]. Whether it be meat that is boiled (or greens [*lakhana*], or something else), it comes to be boiled whenever the moist in it, being undetermined, is both concocted and (*te kai*) determined by the heat in the external moist.

25   The name 'boiling' is properly applied, says Aristotle, only to those things which are boiled. He made clear what the inherent, undetermined [part] in the moist is by saying, 'This is, as has been said, airy or watery.'[76] Having stated that boiling occurs by means of moist heat, in explaining what moist heat

30   is he says; 'Concoction occurs' (that of the undetermined airy or watery [part] in the moist, of course) 'from the fire in the moist', the equivalent of saying, 'from the hot in the moist'.

Aristotle has discussed boiling and given the account of boiling. Since fried things seem to be fried in moist hot (for these things are changed when the oil in the pan, or whatever other moist substance things are fried in, is heated, and they

35   are not said to be boiled, but fried), he shows that by frying they are not affected (*paskhei*) by the hot in the moist, like boiled things, but by the fire outside (*exô*), like roasted things. Hence

things that are fried are roasted rather than boiled, for fried   192,1
things are changed and affected by the outside hot, that is, by
the fire, not by the hot in the moist surrounding them. A sign
that they are not affected by [the hot in the moist surrounding
them] is that the moist surrounding them is affected more by
the things frying. For the things frying, heated by the fire of   5
the hot iron, dry the moist surrounding them by absorbing it
into themselves.

Boiled things do (*poiei*) the contrary of this. Not only do they
not absorb into themselves the moist hot surrounding them,
but, being affected, changed and heated by it, they also eject
the moisture inherent in them when raw. A sign of this, he says,   10
is that boiled things are drier than roasted things. Because they
are mastered by the hot in the moist surrounding them, things
that are boiled do not draw [the moist hot surrounding them]
into themselves, but discharge (*aphiêsi*) the [moist] inherent in
them. Things that are fried (and generally things roasted[77]
because of the heat generated in them by the [external] fire
when it masters the external moist) draw this [moist] into   15
themselves. That things roasted are more moist than those
boiled is clear from the fact that when they are cut, roasted
things are found to be much more moist than boiled things, if
the roasted things were not pre-boiled.

Having said that boiling is a concoction of the inherent,
undetermined, [part] in the moist, generated by a moist heat,
he asserts reasonably that not every body can be boiled. It is   20
not possible for anything in which there is no moist (for this is
concocted in boiling) nor for anything in which there is [moist],
but moist which because of its density (*puknotêta*) cannot be
mastered and ejected by the heat in the moist surrounding it.[78]
Hence neither stones nor wood are boiled, the former because
they do not have moisture, the latter because they have moisture of the sort that cannot be mastered and ejected by the
boiling moist-hot because it has been naturally made dense   25
(*pepuknôsthai*) in structure and does not ooze (*exienai*).[79] Those
bodies alone boil which have affectable moisture, [affectable]
by the heat in the surrounding moist, which itself is generated
by fire.

Other things are said to boil, Aristotle says, not strictly, but
metaphorically. Gold, wood, and certain other things are said   30

[to boil], not in the same sense (*kata tauton eidos*) [as those which do strictly speaking boil] but metaphorically, because not all differences have proper names. It is not because the moisture in gold is mastered by the heat in the moist that gold is said to boil, nor in the case of wood either. For Aristotle said previously that such bodies do not boil, nor are they placed in the hot moist so that the moisture in them might be ejected
35 when they are mastered by the hot-moist.[80] Rather these bodies
193,1 are said to boil metaphorically because of a certain resemblance [to things which do strictly speaking boil] when, because they are heated by fire, some of the moist in them is ejected. In the case of gold, just as with silver or bronze or anything else, when this ejection of the foreign moisture mixed with it (*enkatamemigmenês*)[81] is produced by the fire, then it is said to
5 boil. So too with other things mined.[82]

Wood, from which smoke is ejected because of the proximity of fire, becomes dry, and is said to have been boiled (it is called smokeless as well).[83] 'Boiling' is applied to moist [things] too, such as milk and must (*gleukous*), and the boiling of these is more similar to what is strictly called 'boiling' than that of gold and such. They [i.e. milk and must] are said to boil when the
10 juice (*khumos*) in them which is itself moist and which is proper to them and not foreign, is concocted and changes into a certain form because of the fire from without [which] surrounds them all around [and] heats them. When the undetermined (*aoristou*) [part of the moist] in them is ejected, whether it is watery, as is the whey (*orrou*) in the case of milk, or airy, as in the case of must, the concoction of the substrate (*hupokeimenou)* and its change to a certain form takes place.
15 'In a certain way',[84] he said, what happens in these cases is similar to what happens in boiling. Because, although they are not placed in hot moist, nor is the moisture in them concocted by it, the aforementioned [e.g. milk and must] when they are in vessels are affected by the external fire [as with things that are boiled].[85]

All things do not have the same end (*telos*), whether boiled
20 or concocted ... (381a1).

All boiled things, he says, do not have the same end and are not boiled for the same purpose (*tou autou kharin*). Some things are boiled so that they become suitable for eating (this is the end of boiling in their case), others become suitable for drinking, and yet others for some other purpose. 'Whether boiled', he said, 'or concocted', either taking these as parallel, since boiling 25 is due to concoction, or, [taking boiling as a species of concoction] since concoction is more general (*koinoteron*) and generic (*genikôteron*) than boiling (for boiling is not the only process due to concoction, but so too are ripening, as he has shown, and roasting, as he will state). He said that all boiled things do not have the same end, and added that neither [do all] concocted things in general [have the same end] since the other forms of concoction do not all have the same end either.[86]

Having stated that not all boiling things have the same end, 30 he supplied as evidence for this the case of drugs, which we also say boil (*hepsein*), although we do not say that they boil in order to be useful for eating or drinking or for nutrition in general. Nor is the boiling of drugs the same as boiling strictly speaking, the definition of which Aristotle gave.[87]

After this Aristotle sums up what boiling things are (speak- 194,1 ing of all boiling in general and not merely that strictly so called). [They are]: 'All things which can become thicker'[88] when heated (for thickening is the concocting of the undetermined moist in them) 'or smaller'; (for when the airy in them 5 is separated out [*diakrinomenou*]',[89] they also become heavier [*barutera*]).[90] For things that become smaller are not necessarily boiled; olive oil, for instance, becomes smaller, but it does not boil, as he will state.

[He said]: 'Some [parts] of them become so, and some [parts] become the contrary'[91] instead of 'some [parts] become thicker, some [parts] thinner', because of the dividing [that takes place] in boiling. Some [parts] of them thicken, and some become thin, as occurs in the case of milk. All that is whey is thinned, while 10 the rest, which Aristotle called rennet (*putian*), is condensed (*sunistatai*) and curdles (*turoutai*) (for it [i.e. the milk] solidifies by having something like [rennet] in it). Foods too might be said to boil in this way, for parts of them are separated into moisture by heat; (urine is [such a separated moisture]), while other [parts] are thickened.

Olive oil, he says, does not boil by itself. It seems to do so 15

when combined with other things, but not by itself, because
when it is heated, it neither thickens nor does it become
heavier,[92] although it does become smaller, nor again does part
of it become more moist, and part thicken. It remains the same
in form during boiling, as long as there is any [preserved].[93]

20 Such, says Aristotle, is concoction with respect to boiling. But
concoction occurs in other cases, not only in the case of boiling.
It makes no difference, he says, whether it occurs by the tools
of a craft (*tekhnikôn organôn*), as for instance boiling brought
about by cooks, or by nature, as in the case of nutrition.
However it occurs, the cause will be the same, for in the
instance of things boiled by the tools of a craft, the hot outside
the things boiling is the cause of the boiling, that is (*te kai*)[94] of
the concoction, of the undetermined moist in the boiling thing.
25 (And [similarly] in the instance of things boiled by nature. For
the nutrition is boiled by the hot outside itself.) The hot which
boils, that is concocts, this [nourishment] is not inside the
nourishment, even if it naturally accompanies the thing being
nourished.

Having spoken of boiling, Aristotle passes to a discussion of
simmering (*molunseôs*), which is the contrary of boiling; and
30 he says that it is the inconcoction contrary to boiling. '[Simmer-
ing] would be the contrary and the primary so-called [inconcoc-
tion]'[95] was written instead of 'the contrary to the primary
so-called boiling', for he has not said what simmering is except
potentially (*dunamei*). For in saying what it is that is called
boiling in the primary and proper sense, he spoke also of its
contrary, simmering, in potentiality. Since boiling was a con-
35 coction of the undetermined [moist] in the boiling body by the
heat in the surrounding moist, the contrary of this would be
195,1 simmering, an inconcoction of the undetermined [moist] in the
body due to a lack of heat in the surrounding moist, as Aristotle
says in his definition [of simmering].

He says that the lack of hot has been said to imply coldness.
He made the statement at the beginning of the book [i.e. *Meteor.*
4.1] in discussing destruction and putrefaction.[96] After saying
5 what putrefaction is he added, 'Hence, since things are affected
because of a lack of hot and everything lacking such a power is
cold, both [the hot and the cold] are causes.'[97] He here reminds
us of this. Simmering and the inconcoction mentioned occur,

he says, 'because of a different motion (*kinêsin*)'.[98] [That is, they
occur] not because the hot produces motion and change, but
because when the cold becomes kinetic, that which causes
concoction, that is the hot, is expelled and driven off relative to
this·moving [cold].[99]

Since he said that the inconcoction which is simmering
occurs because of a lack of heat in the surrounding moist, he
adds a discussion of how it is that [simmering] comes about
because of a lack of hot in the surrounding moist: 'either
because of the amount of coldness in the moist',[100] (i.e. when
the moist surrounding the boiling thing is more cold than it is
hot), or because of the amount of the moisture in the thing being
boiled, when the hot in the surrounding moist is not in such a
ratio to it [i.e. to the moisture in the thing being boiled] as to
master it. For [simmering] can happen in either way.[101] Hence
the heat in the moist surrounding the boiling thing is on the
one hand too much not to move the moist in the boiling thing,
and too little on the other hand to concoct it completely (*sum-
pepsai*) and [that is] to bring it to a uniform temperature
(*homalunai*). When this is the case, and there is such a lack of
hot, simmering occurs. This is when the moist in the boiling
thing is moved, but is not concocted, nor brought to a uniform
temperature.

Because of this unevenness [of temperature], says Aristotle,
things that have been simmered become harder (*sklêrotera*)
than those that have been boiled. In the same way that things
that have been simmered become harder than those that have
been boiled because of this unevenness, so too in the case of
moist things, which also are said to be boiled; like milk they
come to be divided because of this unevenness, condensing
(*sunestôta*) in part and in part flowing about.

Having said that simmering is the inconcoction contrary to
boiling, and adding that, '[simmering] would be the contrary
and the primary so-called [inconcoction]',[102] he did not further
state what other [contraries there are] because simmering
properly so called is the above-mentioned [inconcoction]. This
is because it would be easy to define the types of simmering
mentioned, treating them as secondary in relation to the first-
mentioned [type]. For they will be the inconcoctions contrary

10

15

20

25

30

to the types of boiling which are so called in a more secondary and general sense.

196,1   Roasting (*optêsis*) is concoction by dry and foreign heat (381a23).

Aristotle has shown what roasting is and how it differs from boiling. Both are kinds of concoctions, which come about by
5   means of foreign heat (not, like ripening, by means of proper [heat]). They differ from one another in so far as boiling is brought about by moist heat, roasting by dry [heat]. For the moisture in roasted things is concocted and determined by the fire itself whether the [thing roasted] lies on the fire directly or indirectly, through some vessel (since the fire is under the vessel which contains no moist; or it does [contain some moist],
10   as in the case of things fried [*tagênizomenôn*], but the [moist] does not master the moisture in the roasted thing; rather it is mastered by it). For Aristotle said that frying is also a form of roasting.

After defining roasting and stating that it is brought about by a foreign hot-dry, he says that because roasting is of such a character (*toiouton*) and comes about in this way, even if in *boiling* something one should make the thing boiled change and
15   the moist in it be concocted not by means of the heat of the moist surrounding it, but by means of the heat of the fire, when the ripening in it has been perfected (*epitelesthêi*), it is roasted, not boiled.[103] This happens when there is little water and the power of the fire overcomes the boiling thing and dries it up (*anaxêrai-*
20   *nousa*). And if the heat from the fire prevails over the things boiled in little water, the boiled things are said to be burnt because they were so affected by the fire. For everything that has become drier after being perfected (*epitelesthênai*) and concocted is so affected [i.e. made drier] by dry heat. Hence, he says, the outer [parts] of roasted things are drier than the inner [parts], because the surface is dried up by the fire because it is near the fire, while the moisture in the middle is not cast out
25   (*ekkrinomenês*) because the outer [parts], which are near the fire, are heated more and [so] contract (*puknousthai*) before-hand.[104]

The reverse is true of boiled things. The outer [parts] are

more moist than the inner [parts] because the outer [parts] are
soaked through (*diugranthai*) by the moist boiling them, but
those inside are dried up (*anexêranthai*) by the ejection of the
moist there. Hence in cases where the process comes about by
art (*kata tekhnên ginomenôn*) it is a greater task, and more 30
difficult, to roast than to boil; for it is difficult in this case to
heat evenly and alike the inner and the outer [parts]. For those
parts that are nearer [the fire] are quickly heated, and, contracting (*sunionta*) and thickening, prevent the inherent moist
in the roasted thing from being ejected. It is shut inside when
the pores, contracting, shut.[105]   197,1

Aristotle compared things produced by art (*kata tekhnên*)
with those produced by nature, since also among things produced naturally, as he will state, some are roasted, others
boiled. For art imitates nature. When, for instance, nutrition
is concocted by the heat in the moist (the very moisture which
is in the parts in which concoction occurs and which is mixed 5
with the nutrition itself), this is natural boiling.[106] But when
the hot itself increases and overcomes the moisture by drying
it up, something like roasting takes place and the nutrition is
said to have been burnt. Hence not only the nutrition mastered
by the cold is unconcocted but also the nutrition that has been
produced by the hot in the aforementioned manner.

He says that in the concoction of nutrition which takes place 10
in the upper belly (*anô koiliai*),[107] no living creature is generated, as some have thought,[108] but [living creatures are gener-
ated] in the post-digestive remains and the excrement, and in
the putrefying [contents of the] lower belly,[109] just as in the case
of things putrefying outside [the body] certain living organisms
are observed to be generated, as he said when discussing
putrefaction. Once they are generated in this manner and in
this place, frequently they then will rise up to the upper belly.
These are what are called bowel and intestinal worms.[110] Hence 15
they are frequently also vomited in sickness. The cause of the
fact that they are generated below, and often rise into the upper
belly, has been stated elsewhere, he says, namely in the *Problems*.[111] [They are generated] down in the lower belly because
they are a result of putrefaction, and that is where putrefaction
occurs. He also states in that work which causes make them
rise to the upper belly in the case of illness.   20

Simmering and the inconcoction produced in that way are contrary to boiling and such-like concoctions. And he says that there is an inconcoction opposed to the concoction called roasting also, but such an affection is harder to name. It is *like* what is called searing in the case of roasted things, which occurs

25 when the roasting is incomplete because of lack of heat. The lack of heat comes about either because of the smallness of the roasting fire, or because of the great amount of water in the thing being roasted.[112] For when the heat from the fire is too great not to move and change the thing being roasted but too little to concoct it, then the inconcoction called searing takes

30 place, being contrary to the concoction that happens in roasting.

'What concoction and inconcoction are then.'[113] Thus Aristotle sums up his remarks. Having mentioned concoction and inconcoction, the general terms which were contrary to each other, and ripening and rawness, which were in turn contrar-

198,1 ies, and having mentioned boiling and roasting, he added to these, 'and their opposites',[114] which were simmering and searing, simmering opposed to boiling and searing to roasting.

CHAPTER 4

5 We must tell of the forms of the passivities, the moist and the dry (381b23).

Aristotle has spoken of the operations of the active powers, the hot and cold, [and said] what they do (*ergazontai*), and how many [things they do] in the generation of substances (*ousiôn*) and in substances already composed (*sunestôsais*). He passes on in turn to the account of the passive powers, which are

10 dryness and moisture, and he investigates what the forms of these are.

He says that the dry and the moist are principles (*arkhas*) of the affections (*pathôn*) in bodies. They are the material principles and [material] causes of the affections in bodies, or else they are the primary affections (*prôta pathê*) in bodies. And the other affections in bodies are compounded (*memikhthai*) out of these, he says, and of whichever of these the bodies have the

greater share, to that affection they are assigned. For those   15
compounds (*miktôn*) having more of the dry are called dry,
those [having more] of the moist, moist.[115] And all the affections
in bodies, he says, will be in them [i.e. in the bodies] in some
cases in actuality (*entelekheiai*), in others in potentiality (*dunamei*). Or, [for] all the bodies, it is either because it is in
actuality that [each] is in the affection that it will be called body
[of a certain sort], either because it is like this in activity, or
because it is able [to be like this in activity].[116] In showing how
actuality and potentiality differ he suggested, 'melting (*têxis*)   20
is related in this way to the meltable'.[117] For 'meltable' is used
in virtue of a thing's being able to be melted. For that is melted
in which melting is already actually present.[118]

Since the moist is easily determined ... (381b29).

Having stated that the dry and the moist are principles, and   25
primary among the affections in bodies, and that all the other
[affections] are compounded out of these, he describes the
relation between the dry and the moist, and how their mixture
(*mixis*) occurs.[119]

Since, he says 'the moist is easily determined'[120] (for, not
being determined by any limit of its own, it is easily determined, always being determined and being conformed to its
surroundings) and 'the dry is hard to determine'[121] (for it has
its own limit and shape and does not conform itself to its
surroundings), they undergo an effect in relation to each other,   30
similar to that [in the case of] a food and its seasonings
(*hêdusmata*). For when salts, and seasonings generally, are
combined with a food, out of both a single thing is created
(*apoteleitai*), when the food takes on an edible character from
the seasonings, by the mixture [of the two]. As in this instance,
so too with the dry and the moist. For in these cases too the dry,   35
by the mixture of the moist with it, acquires the ability to be
determined, and in being mixed with one another, each be-   199,1
comes to the other a sort of glue. For the moist causes the dry
to be congealed (*sunestanai*) and not to crumble (*diapiptein*),
and the dry likewise causes the moist not to dissipate or scatter,
but to have consistency (*sustasin*).

After stating that the dry and the moist become for each

5   other a sort of glue when in mixture (*mixei*), Aristotle men-
    tioned that Empedocles said something like this with his
    'gluing meal with water'.[122] He too compared the mixture of the
    moist and the dry with gluing.

    Hence Aristotle says that every body both composed and (*te
    kai*) determined[123] is [composed] out of both moist and dry
    because each of these is undetermined if taken unmixed (*ak-
10 rata*).[124] He also says which of the simple bodies are those
    especially given form (*eidopoioumena*) by these powers, namely
    earth by the dry and water by the moist. Since all determined
    bodies, as he said, are [composed] out of (*ek*) a mixture of moist
    and dry, and moist and dry [correspond with] water and
    earth,[125] all determined bodies are [composed] out of water and
    earth. Aristotle added the word 'here'[126] (*entautha*) which
15 makes clear [that he meant] 'in bodies subject to generation
    and destruction', for in these cases, the determined [things] are
    necessarily [composed] out of both of these [powers]. For this
    is not the case with divine things.[127]

    When these [two powers] are mixed and determined, which-
    ever of them is in greater quantity in the mixture and which-
    ever power predominates (*huperekhêi*), [the mixture] is called
    that. Stones are earthen (*gês*) because they contain the dry and
    earthy in greater quantity, while honey, wine, and such juices
20 are called watery. Hence, he says, creatures exist and are
    generated in (*en*) earth and water alone, but never in (*en*) fire
    or air, since living things must have a material substrate
    (*hupokeisthai*) and matter is 'the passive', and since the moist
    and the dry (which are water and earth) are passive.[128] That
    winged creatures are also generated in (*en*) earth is not difficult
    to understand.

25   These must be the first of bodily affections to come into
        being ... (382a8-9).

    After showing that every determined body has composition
    from the dry and the moist, Aristotle adds that every thing that
    is determined (that is to say, composed out of [*ek ... sunke-
    imenon*] dry and moist) must have hardness or softness before
30 any other affection. That which has more of the moist is soft,
    and that which has more of the dry is hard.

Having stated that each determined body must be either hard or soft, he says what hardness and softness are. He says that a hard thing is that which does not give way into itself at its surface, that is, which does not yield at its surface, but withstands and resists. 'A soft thing is that which gives way by not being displaced (*antiperiistasthai*)',[129] that is to say, which yields by contracting (*sustellesthai*) at its surface to the touch, but without being displaced when it is divided (*diairoumenon*). For water and air, although yielding to the touch, are not soft. For they [water and air] yield by being displaced when they are divided, but not by their surface, while remaining continuous, contracting inward where squeezed.

200,1

5

He has defined the hard and the soft. Since some compound bodies are and are called hard or soft without qualification (*haplôs*), and others are relatively so (the same thing is hard compared with one body and soft compared with another), he says that those bodies which are soft or hard without qualification (rather than by comparison with something else) are the things which admit of the aforementioned accounts. Those which are hard relative to other things or soft relative to other [bodies] are those that are like this [that is, like the hard or soft without qualification] relative to those [other bodies] and admit of these accounts [that of being hard or soft] more than those [other bodies] do.

10

Bodies which are relatively hard are such as to give way at their surfaces less than those than which they themselves are harder, and, conversely, relatively soft bodies are those that give way more [than those things than which they themselves are softer] by contracting inward from the outside.

15

Things which are said to be relatively hard or soft rather than hard or soft without qualification are indeterminate (*aorista*). The indeterminateness in them is a matter of degree. That which is simply hard or soft is not indeterminate, otherwise it would not be [hard or soft] without qualification.[130]

Aristotle next tells us with what the unqualified hard and unqualified soft are compared, and by what they are determined. For, since the proper sense is a standard of all perceptibles, clearly touch is the measure of hard and soft. It therefore determines the unqualified hard and unqualified soft. Aristotle says how they are determined: since touch is a mean (*en*

20

mesotêti) between the objects it perceives (for touch does not perceive [*aisthanetai*] tangible things [*tôn haptôn*] in the same
25   way that sight perceives colours; for sight is capable of apprehending all colours because it does not itself have a proper colour and because it receives the forms of them [colours],[131] but touch does not apprehend its objects in this same way, through itself being free from tangible affections, for it is impossible for a body to exist actually without some tangible affections, for they are what determines the body, and it is a body through which touch operates, so that this [body] too is
30   tangible), not then because it does not have these affections, but because it is a mean between them, participating in both the contraries equally, that is [being] mixed from an excess and deficiency of these, by means of this median status it comprehends (*antilambanetai*) [both]. Hence touch does not perceive all tangibles as sight does all colours. Touch cannot apprehend
201,1   that which is as hot as itself, or as cold, hard, soft, or moist, as itself, but can apprehend what exceeds it or falls short of it with respect to each of these [affections].[132] And so it stands with hardness and softness.

So that which exceeds the mean which is touch in respect to hardness will be hard without qualification and will be appre-
5   hended by touch as unyielding (*mê hupeikon*). Conversely, that is soft without qualification which the sense [touch] lays hold of (*antilambanetai*) because it yields to it in a greater degree. Hence that will be hard without qualification which is harder than the touch which lays hold of it, and that will be soft without qualification which is softer than the touch which lays hold of it. And in this way touch is a mean (*mesotês*).[133]

The senses are called means in so far as they can receive each
10   of two contrary forms, for that which can receive contraries is in some way a mean [*meson*] between them. They are also means in so far as they are destroyed (*phtheirontai*) by deficiencies and excesses of the perceived [affections] and are preserved by moderation and a median status [*meson*].

'We say the delicate (*lepton*) is also soft.'[134] Aristotle wrote 'delicate' for 'flexible' (*endidon*). For the delicate is flexible just
15   as the thick is resistant. He also gives the cause of why every

determined body is hard or soft: it must either yield to the touch or not; and of these the former is soft, the latter hard.

CHAPTER 5

Further it is solidified (*pepêgos*) ...[135] (382a23).

Having taken every determined body to be either hard or soft, he adds another affection to it; he says it is necessary for a 20 determined body to be solidified. For the hard is hard and the soft is soft in virtue of a certain solidity (*pêxei*). Every body both determined and (*te kai*) composed, whether it is hard or soft, is such as it is in virtue of a certain sort of solidity; no body is compounded (*suntheton*) and determined without solidity. Hence in speaking of bodily affections (*sômatikôn pathôn*) one 25 must discuss solidity.

In his discussion of solidity he takes up its causes. He says that there are two causes besides matter, 'the agent (*poioun*)[136] and the affection (*pathos*)', meaning by affection the form coming to be (*ginomenon*) in the matter by means of the agent. There are then three causes, matter, form, and the agent, because the end [i.e. final cause] is reckoned in with the form. For form seems to be the end in things that come to be naturally.

If just these are the causes in all cases of coming-to-be, and 30 if solidity is [a feature] of things that come to be, then these are also the causes of solidity, and of dissolution (*diakhuseôs*) as 202,1 well (which is the contrary of solidification). [They are] also [the causes] of drying and of moistening, by means of which solidification and dissolution [occur], as he will remark.

Aristotle says that the agent acts by using certain powers, that is, the hot and the cold. These are the active powers because through them the agent acts (it acts by warming or 5 cooling). If the agent acts thus through these powers, then the affection produced in the subject (*paskhonti*) is produced because of the presence of hot or cold.

Since solidification is in a way a drying ... (382b1).

Aristotle has spoken of hardness and softness, which he said
10   were the primary affections of bodies both compounded and (*te
kai*) determined. In proposing to speak next about solidity (for
this affection is produced in bodies by the active powers), [and]
before saying what solidity is, he speaks of drying, since solidification is produced by drying. For that which is made solid is
dried in a way.
15   He says that the subject is either moist, dry, or composed out
of both,[137] for these are the passive powers. It was laid down
that water is [constituted] of the moist, earth [constituted] of
the dry, that is to say, water is given form by the moist, earth
by the dry. Hence, of the [simple] bodies containing in themselves moisture or dryness, these [i.e. water and earth] are the
passive ones. For fire too contains dryness and air moisture,
20   but they are not passive because they are not given form by
those powers.[138]
Since water and earth are passive, and since coldness is in
both (earth contains the cold in addition to the dry, water is
cold as well as moist), for this reason the cold also seems to be
one of the passive powers rather than one of the active. Since
25   it was laid down that the cold was active[139] but he has just now
said that it is passive [rather than active] because it is found
in passive bodies, he says in what sense it is active, namely, 'as
a destructive agent'.[140] For since destruction too is an action,
and things destroyed are destroyed by an excess and predomi-
nance of coldness, the cold is active in this way. 'Or incidentally'[141] (*kata sumbebêkos*), when anything is generated by the
hot either because the hot inherent in something which is
30   chilled is concentrated[142] as the cold gathers (*sunagontos*) and
contracts (*puknountos*) it (as happens with thunder and lightning;[143] but concoctions also occur in this way), or because the
hot, displaced by the cold, migrates and withdraws (for instance, the earth as well as that under the surface becomes
warmer in the winter; hence well-springs [*pêgai*] are warmer
then[144]). For the cold seems to be the cause of these phenomena,
203,1 not because it acts directly (*prosekhôs*), but either because the
hot is rounded up (*êthroisthai*) and becomes so great [in con-

centration] that it can act, or because the hot has the cold as the reason for its being in certain places.

Whatever is water and the forms of water are dried ... (382b10-11).

Intending to speak about drying, but first reminding us which 5
bodies are passive (they are either moist[145] or dry or mixed
(*mikta*) out of these), Aristotle now says what the things that
are dried [are]. He says that these are dried: 'whatever is water
and the forms of water.'[146] He will shortly say what the forms
of water are. They are those liquids which have little or no
sediment, such as wine, urine and whey. These are dried, and 10
also those things [are dried] which contain water, whether they
acquire this water from without and it is brought in, and not
in their proper nature, as in wet fleeces and garments (for what
it is for a fleece to be a fleece does not consist in its being wetted,
but the water which is in it when it is wetted is outside its [i.e.
the fleece's] own substance), or they have it in their proper
nature. For those things containing water in this way are dried 15
too. Such is milk, for it has the moist in its substance and
nature; for it is in this that being milk consists.

Having said what the forms of water are and how many there
are, he says that they are 'all things that have little or no
sediment, [but] not on account of [their] stickiness (*glis-
khrotêta*)'.[147] [He says this because] sticky [fluids] do not suffer
this [affection; i.e. that of not having much sediment] not
because they do not contain much potential sediment, but 20
because of their stickiness. [For] when the sediments are mixed
in them in this way – so that they cannot be loosened (*dialuesthai*) and separated from the moist in the sticky things – [then
these things are sticky and do not have much sediment].[148]

Aristotle has told us what the things that are dried are, and
next tells how and by what they are dried. Now, the things that
are dried are all dried either by heating or cooling. But, he says,
even the things which are dried by being cooled are dried by 25
hot, for when they are cooled the hot in them, when it is
evaporated[149] (*exatmizomenon*) by the cold, carries away with
it (*sunexagei*) the moist which is in them (if there be little of it);
thus they are dried.

Having said [that drying occurs] by means of either the internal or the external heat, Aristotle showed how by the internal heat certain things are dried. He adduced the case of garments drying in the cold when they have been soaked

30 [*bebregmena*] and [so] have in them an independent (*auto kath' hauto*) and separate moist.

Now there are some things which have a connate (*sumphoton*) moist (as, for example, milk) which are not dried in that way, [that is, they are dried] by the heat *in them* when it evaporates because of the cold, and evaporates their moisture along with it. Such things are not dried in *this* way, rather they

204,1 are dried when the moist in them is evaporated by the *external* hot.

Everything, then, which is dried is dried either when it is cooled or when it is heated and everything [is dried] by means of hot, as has been shown, whether by the external [hot] when it heats and evaporates the moist in drying things, or by the internal [hot], which, when it evaporates out of the drying

5 things because of the cold, evaporates and carries away along with it the moist in the drying thing.

Things which are dried by external heat are dried in the same way as things boiled,[150] while those dried by internal heat are dried in the same way as boiled things, when the external fire is taken away, and the moist within them is consumed by the heat in them, which is blown out (*apopneousês*), carrying the moist away with it. For when things are cooled the hot is

10 blown out and the moist is carried away with it.

Aristotle has stated what things are dried, how [they dry], and by what [they are dried] (for moist [things] are dried, he said, by being heated or cooled, when the moist is consumed and destroyed by the external or internal heat).

CHAPTER 6

Now Aristotle passes on to an account of liquefied things. This

15 [i.e. liquefaction] is an affection contrary to being dried. He says that liquefaction occurs in two ways: whenever a change into water from a condensed (*sunistamenês*) and thickened exhalation (*anathumiaseôs*) comes about (for it liquefies when it

changes into water), or whenever a solid (and so concrete) [thing] melts. Such a thing is liquefied too. He says that what is condensed (*sunistamenon*) and liquefied in this way is pneuma, meaning by 'pneuma' the exhalation, which although 20 vaporous (*atmidôdês*) has something of the smoky (*kapnôdous*), dry exhalation [in it] too. That [exhalation] in which the vapour predominates is condensed (*sunistatai*). [What] melting [is], and [what] those things which liquefy through melting [are], will be [made] clear when we speak about solidification, says Aristotle.

He speaks about solidification, and first he takes up the question of which things are solidified. He says that whatever 25 is watery, or whatever is of earth and water, is solidified. Such are clay, wax, and similar things. After saying which things are solidified, Aristotle next explains by what and how [solidification takes place]. He says that solidified things are solidified by hot or by dry cold.[151] A sign of this is that things which have been solidified are dissolved by causes contrary to those causes which solidified them (those, at any rate, of solidified things 30 which can be dissolved), and some solidified things are dissolved by water, which is cold and moist, because they were solidified by what was hot and dry. That things solidified by fire become solid not in so far as fire is hot (i.e. they are not solidified by fire in so far as it is hot alone), but in so far as it is also dry is clear from the fact that things are not solidified by hot water because it is not dry as well as hot.[152]

It is clear that things which are dissolved by fire were 35 solidified by cold. But things dissolved by fire (since fire is hot 205,1 as well as dry) could not have been solidified by moist as well as cold, for the moist is not an active cause (*aition hôs poiêtikon*) of solidification for anything, since it is itself the matter (*hôs hulê*) of solidified things. Aristotle shows that the moist is not an active [cause] of solidification by showing that cold water is 5 not the cause of solidification in the boiling of honey in so far as it is moist but only in so far as it is cold. For honey is solidified by what is cold and not moist, but not by [what is] moist and not cold, just as a solidified thing is solidified by cold-moist (that is, by water), not insofar as it is moist but insofar as it is cold.[153]

'Whatsoever is of water ...' (383a6-7). Since Aristotle has

stated that some things are solidified by dry-hot, others by
10 cold,[154] he says which things are solidified by each of these.
Some, then, because they are of water (*hudatos*), are not solidified by fire. For whatever there is that is [composed] out of
water, whenever it is solidified, that is dissolved by [fire]. But
it is impossible that the same thing [acting on] the same thing,
by means of the same thing (that is, by the same power), should
be the cause of contrary effects in the same thing. Hence, if fire
dissolves solidified things [composed] out of water by means of
15 heat, such as ice, then it could not solidify [them] at the same
time by means of [its] heat.[155] Nor is it possible for the dryness
in the fire to cause the solidification of such things. For they
would be dissolved by moist [if the dryness in the fire were the
cause of their solidification], but actually neither of these [i.e.
the dry and the moist] is capable of dissolving [anything] unless
it were accompanied with heat, because it is by the hot that
things dissolve.[156]

Further, watery things are solidified because the hot in them
20 leaves, and so they will be dissolved [by its] entrance. Supposing that it was evident from his first attempt that this kind of
solidified thing is dissolved by fire, Aristotle showed that it is
not solidified by fire. Supposing [that it was evident] from his
second attempt *how* such things are solidified, he constructed
an argument that they cannot be solidified by fire (on the
contrary they are dissolved by it).

Having shown that it is not by fire that whatever is [com-
25 posed] out of water has been solidified, Aristotle says by what
they are solidified, namely by cold. Because of this he says
things [composed] out of water are not thickened when they are
solidified. For thickening occurs in solidified things by means
of the moist departing from them and the dry in them being
compacted (*sunistamenou*). But it is impossible for water and
whatever is a form of water to have thickened, because they
have the earthy in their own nature either not at all or only a
very little.

30 Having stated that solidified things [composed] out of water
are not solidified by fire but by cold, Aristotle says which things
are solidified by fire: those mixed (*mikta*) of earth and water.
206,1 These, he says, are solidified by fire *and* by cold. Such mixtures
of earth and water are thickened, he says, for the solidification

(*hê pêxis*) of these (not of all, as he will mention, but of these) is by thickening. He says that they are thickened by both [i.e. fire and cold], in some sense in the same way, in some sense in a different way. [They are thickened] in the same way in so far as when it occurs that the moist from them is drawn out by the 5 hot it thickens them in both cases, but [they are thickened in] different ways in that the manner in which the moist is separated (*khôrizetai*) from them differs. For the hot, just by heating the solidifying thing from without, evaporates the moist out of it and thickens it. The cold, however, does not separate out the moist in this way, but squeezes out (*ekthlibei*) the hot in the 10 solidifying thing, and because the hot is squeezed out the moist is carried along as it evaporates and is thus also separated out by means of the hot.

Aristotle next says which solidified things first thicken [when they are solidified]. He says that of the things mixed out of water and earth, some are soft, others are moist, meaning by 'soft' that which he goes on to call softened bodies (such as clay), and by 'moist' such things as milk. Of these, he says, the soft do not thicken when roasted, but the moist thicken first 15 [before they are solidified].[157] Pottery, for example, when it is roasted does not thicken at first when it dries, but whatsoever is moist and mixed[158] thickens [before solidifying], as does milk when the whey is being separated.

He said, 'whatsoever of *mixtures* (*mikta*) are moist',[159] and rightly, for water is moist, but because it is not a mixture, it is not thickened. For he has just said that water alone of moist 20 things does not thicken. Many soft things, he says, which were originally (*proüpêrkhen onta*) thicker or harder through being solidified by cold, not only do not thicken in roasting, but first become moist. For these, when heated, first are dissolved and remoistened, then dried. And he said that pottery is such a thing. For when pottery is roasted, it first gives off vapour as the solidified moisture in it is dissolved, and [so] it becomes 25 softer. This causes it to become misshapen, for it is when it becomes softer that it becomes misshapen. For in coming to be softer it is distorted, for what is softer is more easily acted upon. The unevenness of the roasting heat is what distorts the pottery.

Of those things, he says, which are composed (*sunkeimenôn*)

30   out of earth and water, and in which the earth predominates, those which are solidified by cold [are dissolved and melted by hot].[160] [They are solidified by cold] when the hot by itself evaporates from them and does not also get rid of (*sunekkrinontos*) the moist in them because, not exiting in a great mass (*athroon exion*), it does not carry away the moist with it. This occurs in those things containing little hot, not so much as to draw the moist [out] and overpower (*biasasthai*) it. Again, he

207,1   says these things are dissolved and melted by hot. For when the hot (the departure of which solidified them) enters into them [i.e. the now solidified things] again, they are dissolved again and resume their former nature. This is what happens to clay solidified by cold; when heated, it becomes moist again,

5   is dissolved, and becomes clay [again].[161]

    All things which are solidified by cooling, because the hot evaporates (*sunexatmisantos*) [and] carries the moist out of them, are indissoluble, says Aristotle, except by extreme heat. For instance, neither chance (*tukhousa*) heat, nor heat from the sun, can dissolve any metal, to such an extent have the metals been solidified, he says. They can, however, be softened, and they are beaten and softened, as are iron and horn.[162] For horn

10   too yields when heated.[163]

    'Wrought iron also melts so that it becomes moist, and solidifies again' (383a32-3). Having remarked that some things are indissoluble except under extreme heat, Aristotle states that these nevertheless are melted by extreme heat. At any rate, wrought (*kateirgasmenos*) iron,[164] while it remains iron,

15   is melted again by exceptional heat so that it becomes moist. These things [are not melted again] through the return of the pre-existing (*proüparkhousês*) moisture because of the presence of the hot, but through the presence of moisture in potentiality in these same things (for their generation was not out of solidified water but out of an exhalation, which was water potentially though not actually) when this is changed into actual [moisture] by means of fire.[165] [Aristotle] says that from

20   the change produced by fire of the iron into a moist [state] a purification (*apokatharsin*) takes place each time. The earthy, which is dross, is separated out and when, after numerous treatments, it is purified, then it becomes steel (*stomômatos*).

The iron purified the most is the best, but workers do not purify it [so much] because of the great amount of loss.

The stone pyrimachus[166] melts too, he says, in a strong fire,　25
and the part of it which flows and drips off becomes hard again when it is solidified. He says that *'mulas'* also melt. But what he means by *'mulas'*, should be questioned for he does *not* mean the stones called *'muliai'*. For he says of them a little later that they are not dissolved.[167] He says the part which flows off (*aporreon*) from the [*mulai*] when solidified becomes black in　30
colour like lime, i.e. unslaked lime. Mud (*pêlon*) also, he says, and earth (meaning earth that has moisture in it, such as that from which glass is made) are melted.

Having discussed things solidified by cold, Aristotle next speaks about things solidified by dry heat. Of these, he says,　208,1
some are altogether insoluble, others are soluble by moist. For example, pottery and certain kinds of stones produced by the burning of earth are insoluble. (He says that the stones called *'mulias'* are like that; a little while ago he said that *mulas* are melted by fire, but perhaps he meant something else by *mulas* there, and not the stones called *'mulias'*.) Nitron and salts are　5
dissolved by cold-moist. Water is such a thing [i.e. a cold-moist thing]. For this reason nitron and salts are melted by water and the forms of water. He mentioned wine, urine and whey as forms of water a little previously. Vinegar may be included with them. But these things are not melted by oil, because the dry-hot, by which [these things] were solidified, [has] as its contrary the cold-moist, which dissolves things solidified in this way, and oil is not cold nor is its moisture piercing, being　10
accompanied by stickiness (*gliskhrotêtos*).

CHAPTER 7

Of the things which are mixed (*mikton*) out of water and earth, those that contain more water than earth, for example, milk, are only thickened by fire and not solidified [by it]. But those which contain more earth than water are solidified [by fire]. Hence nitron[168] and salts are more of earth, for they contain more of it and so are solidified by fire. So it is with pottery and　15 any stones which are generated when earth is burned by fire, as he has said.

There is a serious difficulty, he says, about the nature of olive oil. Does it contain more water or earth? For if it contained more water than earth, it ought to have been solidified by cold, and if it contained [more] earth, by hot. But in fact it is solidified
20 by neither, and it is thickened by both the cold and the hot. The cause of this, he says, is that olive oil is full of air. As a sign that olive oil is full of air he cited the fact that it floats on the surface of water. For the air in it is the cause of this, since it floats on the surface of water, and rises up, should it be trapped (*enapolêphtheiê*) in water.
25 He next tells us that the air in olive oil is the cause of the fact that it is thickened by both the cold and the hot. For the cold, by chilling the pneuma, that is the air, and changing it into water, thickens it. The result is that when the air in the olive oil too is both condensed (*sunistamenos*) in the process of being chilled and (*te kai*) also is thickened and changed into water, this causes the olive oil too to thicken.[169] For whenever
30 olive oil and water are mixed, their mixture is thicker than either the water or the olive oil unmixed with water. It must not be supposed that all the air in olive oil is turned into water (*exudatousthai*) by the cold, but only part of it; for if it all were, that which remained would not be olive oil if what it is to be olive oil is to be full of air.[170]

On the other hand, he says olive oil is thickened and lightened [in colour] (*leukainesthai*) by fire and also by the lapse of
35 time. It is lightened because the hot evaporates any water in it, the admixture of which made the colour of the oil darker,
209,1 and when the water is evaporated, the oil is thickened. Fire is thus the cause of its thickening and lightening. Time causes olive oil to thicken but not to lighten, for when the air in the olive oil diminishes and the hot is cooled, the air, changing into
5 water, makes the olive oil thicker.[171] 'The same affection (*pathos*) arises in both cases',[172] both by means of fire and by means of time, or '[the same affection arises in] both cases' because the water is evaporated and because the air in the olive oil changes into water.[173] Through both of these the same affection comes to be in olive oil, (it is thickened) and by means of the same thing (for it is thickened by means of water), but not in

the same way; rather, it is sometimes by its evaporation and 10
sometimes by its generation (*gennasthai*).

The words 'in both cases' and 'by means of the same thing'
may also have been said applying to both cold and the passage
of time: olive oil is thickened by both because water is generated
in the oil out of the change in the air, but the cold chills it
directly, while time weakens the hot in it. Olive oil is thus 15
thickened by both, but is not dried up by either, because neither
air nor sticky moist things are dried either by the sun or by
cold. For stickiness prevents drying and air is not a thing that
is dried, but, as [Aristotle] has said, water and the forms of
water [are dried]. For these are dried by evaporation or by
solidifying. But the water in olive oil is neither boiled by fire 20
nor dried [by it], because water dried by fire is dried by both
vaporising (*atmizein*) and changing into air, but the water in
the olive oil does not vaporise, nor does it boil, because of the
stickiness of the olive oil.

He says that all things mixed out of earth and water must
be classed with the thing of which they contain more, with 25
water if the water is more, or with earth if the earth is more.
He says that a certain wine is both solidified and boiled (that
is that it is thickened). Now solidifying applies to those things
containing more of the dry, and thickening to those containing
more water. By this he will appear to say that a certain wine
contains more earth than water, since it solidifies. Yet he has
said that wine is a form of water. Or rather, it is not simply 30
that it is solidified that is a sign that this [wine] contains more
earth than water, but that it is solidified by fire, since water
too is solidified, [but] by cold.[174] Thus wine which is solidified
is also solidified by cold. But must (*gleukos*) is thickened when
boiled (although water is not thickened [when boiled]), because
it is not yet a pure and separated[175] wine, but is still unconcocted.

The water, he says, departs (*aperkhesthai*) from 'all such 210,1
things' as they dry[176] – i.e. things mixed out of earth and water,
and having more earth, which are dried by hot. For when they
are dried, the water departs. The vapour (*atmis*) rising from
them when they are drying is water,[177] as a sign of which he
adduced 'the fact' that the vapour, when collected, is condensed 5

(*sunistasthai*) into water. At any rate the vapour which is collected on the lids of kettles is water.[178]

In the case of things evaporated by fire, whenever there remains something which cannot be evaporated, this [residue] one must suppose to be of earth. For in [the case of] water, whenever it is vaporised and dried, if it is pure water (*kath' hauto on*) it leaves behind nothing which cannot be evaporated.

10 He says some of the things mixed out of earth and water are solidified, when they are both thickened and dried by cold, for the cold does not only solidify and dry [things], but also thickens [them]. It solidifies and dries water, for that which is solidified is also dried. He stated that solidification too is a kind of drying.[179] [Cold] also thickens oil, changing the air [in it] into

15 water by chilling; indeed it thickens air itself in the change to water and formation of clouds (*nephôn*).[180]

Because he had said '[cold] dries water', he followed with, 'We called solidification a certain type of drying',[181] for the cold dries the water by solidifying it in this way. Having said that some things are solidified and dried by cold, and that some also are thickened, he says that whatever is not thickened by cold

20 but is solidified [by it] is watery, such as wine, urine, vinegar, lye and whey. Whatever is thickened by cold, but is not thickened by evaporation caused by fire, as milk is thickened – whatever is not like milk in this respect, is either preponderantly of earth, he says, or of both water and air. Honey is preponderantly of earth, for honey is thickened by cold and not only solidified by it. Olive oil is clearly of air and water. (This

25 Aristotle neglected to add.)

He says that milk and blood are out of both water and earth, with rather more earth.[182] Most blood and milk are *generally* so [composed], for milk or blood can become watery by having more of water. The fluids (*ta hugra*) too out of which salts and

30 nitron are solidified have more earth, he says.[183] Blood and milk are like them [i.e. these fluids], he says. Certain solids (*sterea*) such as bones, sinews and flesh, come to be out of blood and milk, just as salt and nitron [come to be] out of fluids.

Some stones too, he says, are generated out of the condensing of certain moistures, as with salts. Such are the waters which, in their flow, produce certain stalactites[184] and stones, and themselves doubtless have a preponderance of earth.

He pointed out that whey is watery, while milk has more 35
earth, when he said that the whey is burned off when milk is
boiled, unless it is separated from the milk (this happens to the
water), but the earthy is both condensed, and thickened and 211,1
solidified. Thus the curds are solidified out of the milk. Since
milk has some earth, Aristotle says it is condensed by fig-
juice[185] too. He mentions a method of boiling introduced by
physicians in which, he says, by mixing fig-juice with the milk
they separate the whey and the curds from each other.[186] The 5
separated whey is not thickened because it is watery, but if
boiled it too is burned away, like water.

Having said a little while ago that generally there is more
earth than water in milk and in blood, Aristotle now says that
milk containing little or no curd is mostly of water and non- 10
nutritive. He shows that blood has more earth by the fact that
blood is solidified by being dried when it is chilled. For it is not
solidified by the cold in the same way as water, but by being
dried. This occurs when the heat in it is separated by the cold,
and withdraws and carries away with it the moisture in the
blood. But blood which is neither solidified nor dried, such as 15
the blood of the stag, he says is mostly of water and is cold. As
a sign that such blood is mostly of water he produced the fact
that it does not contain fibres (for they are solid bodies [*stereai*]
and of earth).[187] At any rate, when the fibres are removed from
solidified blood which contains fibres, that which is left behind
is no longer solidified or dried, for what remains is water, just 20
as [the water] of milk [is left] when the curd is separated, for
the whey is water. A sign of the fact that blood which is not
both solidified and condensed is watery is that diseased things
are not solidified and are like serum[188] and blood that has been
wasted and broken down.[189] Having said that diseased blood
because it resembles serum (*ikhôroeidê*) does not solidify, Aristotle says that serum is some kind of phlegm and water. Such
blood is produced because it is not concocted and mastered by 25
the nature of the concocting thing.[190]

Aristotle next offers an explanation of the fact that some
solidified things are soluble, such as nitron and salt, while
others are insoluble, such as pottery, and [an explanation of
the fact that] some of the insoluble things can be softened, such
as horn, and others cannot, such as pottery and stone. He says 30

that since each of the soluble things is dissolved by the contrary of that which solidified it (for whatever is solidified by the cold is dissolved by the hot; and whatever is solidified by the hot is dissolved by the cold), those which are solidified by both causes, both the hot and the cold, are naturally the most insoluble, since they do not have[191] a power to dissolve them contrary to

35　that which solidified them. Whatever has been previously heated and then solidified by cold is solidified by both. It happens in these cases when the hot, leaving [a heated thing], exudes (*exikmasêi*) the moist, then, as it cools, it is compressed (*sunthlibesthai*) again and is contracted (*puknousthai*) in such

212,1　a way that it gives no passage for any moist [to pass through]. Hence these things are dissolved neither by hot (for that which is solidified by cold alone is dissolved by the hot), nor by water (for that solidified by cold is not dissolved by water, since whatever is solidified by dry-hot alone, *is* dissolved by water, as has been said).[192]

5　Iron, when it has first been melted by the hot, is [then] solidified by the cold, and so both [the hot and the cold] are required for solidification in this case. Hence it cannot be dissolved except by excessive (*huperballontos*) hot, as he has already said. Or he means that iron is not solidified by hot, for it becomes more moist as fire melts it, but is solidified by the cold. And so if it is solidified by cold alone, it makes sense that it is dissolved by hot.[193]

10　He says of wood that it is a mixture (*mikta*) out of earth and air. It is therefore combustible (*kausta*), he says, but cannot be melted or softened, whereas things having water are such [as to melt or soften]. That is why wood floats on water, he says, except for black ebony. For black ebony has a preponderance of earth, because the air has been dispersed as vapour (*diapepneukenai*) out of it, while the other kinds of wood have more air than earth.

15　Pottery has earth only, he says. It is not meltable (*atêkton*), for earth which has little or no moisture in it is not meltable. Pottery is such because when it is dried by the fire it is solidified gradually. Because its solidification comes about gradually and not all at once (*athroan*), the water, when it is evaporated, separates from the pottery because it (*hôsper*) changes into air.

20　Hence pottery admits no water by which it might then be

dissolved, because it is too dense. This is because the passages by which air is expelled (*exekrithê*) are not a passage for water, for air is finer (*leptoteros*) than water. This is why pottery is not dissolved by water. And it is not dissolved by fire because it was solidified by it in the beginning.

It has been said what solidification and melting are, how many (*posa*) things cause them to occur, and in how many (*posois*) things they occur (384b22-3). 25

Aristotle again reminds us of what he has said. He has not given a definition either of solidification or of melting, but from what has been said by him each can be defined. Solidification is a condensing (*sustasis*) of water, or of earth and water, brought about by cold or dry-hot, and melting is a dissolution of the solidified thing brought about by the contrary of that 30 which produced solidification. 'How many things cause them to occur, and in how many things they occur.' He has said that solidification and melting come about by the cold, or by the hot and dry, and that they occur in water or in a mixture [composed] out of water and earth. For solidification [and melting] occur in these [that is, water or a mixture of water and earth].

CHAPTER 8

It is clear from what has been said, says Aristotle, that composition occurs in bodies by means of hot and cold. For the hot 35 and cold, as has been said, produce the composition of bodies 213,1 either by drying them or by thickening and solidifying them. And it is because these are the agents producing composition in bodies that there is heat in all mixed and composed (that is determined) bodies, and cold also in some, in so far as the hot in them is lacking.[194] For cold seems to be produced in bodies by a deficiency of the hot, and it is an agent of destruction. 5

Since then, the hot and the cold are inherent in determined bodies because they are the agents in them, while moist and dry [are inherent in them] because they are the passive elements (*paskhonta*) [in them], mixed and determined bodies will share in all of these. Aristotle says that 'homeomers in both

10   animals and plants are composed out of (*ek*) water and earth'[195]
      (these are bone, sinew, flesh, wood, and bark), not because the
      homeomers are composed out of these alone, but because they
      contain more water and earth [than fire and air]. Metals also,
      he says, are composed out of these. They are gold, silver, copper,
      iron, lead, tin, and so forth. The composition in these cases is
15   produced out of water and earth, and, at the same time, 'out of
      the exhalation of each',[196] meaning by 'the exhalation of each'
      the exhalation from water (which is vaporous), and that from
      earth (which is dry and smoky). Out of this dual exhalation
      which is derived from both of these, when it is shut up in the
      earth and solidified, come the metals and things that are
20   quarried. The things that are quarried [are produced] out of
      the exhalation which has more dry, the metals out of the
      exhalation which has more moist. Aristotle has spoken of these
      at the end of Book 3 of the *Meteorology*.[197]
          He says that those things composed out of the aforemen-
      tioned [i.e. water, earth, and the exhalations from these] differ
      from one other, both by properties peculiar to the several senses
      – these are active features – and by their capacity to be affected
25   by certain things. The [features] by which they affect the senses
      are different in them (these are colours, flavours, odours good
      and bad, and different noises – some of these are capable of
      making a sound, some less so – and also tangible contraries);
      they thus differ from one another in these respects, which are
      active, and (he tells us) in passivities peculiar to themselves
      besides; and it is by these that they are given form. These are
30   the features by which they are described, from their capacity
      to be affected by them.[198] For instance, some are said to be
      solidifiable (*pêkta*) (those that are able to undergo this effect
      [i.e. solidification]), others again are said to be meltable (those
      that can be melted), or likewise bendable, and so forth.
          The aforementioned bodies, that is the homeomers, differ
      from one another especially with respect to these affections, for
      they are passive like the dry and the moist. Just as these
35   powers are passive, so are the bodies composed out of them,
      which are precisely the homeomers. For these are composed
      out of water and earth, the water being moist and the earth
      dry. These are natural bodies and the first of compounds

(*sunthetôn*). For anomeomerous [bodies are produced] from the compounding of homeomerous [bodies].[199]

Intending to speak of the affections by which compounds of 214,1 homeomerous bodies are given form (and in turn of the [compounds] given form by those affections), Aristotle first set out what the form-giving affections are and how many in number they are according to [the homeomerous bodies'] ability or inability to undergo any of these affections by means of which compound and homeomerous bodies are differentiated from 5 each other. Advancing eighteen antitheses in which he says that most bodies differ from one other, he next says what each thing is in the antitheses which he set forth, and what power each has.

Since the first two of these antitheses were 'solidifiable and not-solidifiable', and 'meltable and not-meltable', and he has already by this time produced an account of them, he says that a general and common account of them has already been given 10 but, nevertheless, says he will now return to them and speak of them. He says: of solidified bodies, some are solidified by hot, others by cold,[200] as has been said. Those [acted on] by the hot [are solidified] by the drying and evaporation of the moist in them [caused] by the hot. In this way nitron and salt are solidified. Those [acted on] by the cold [are solidified] by the squeezing out of the hot which is in the things being solidified. 15 Thus ice is formed from water when it is solidified because the hot in it is mastered and ejected by the cold. Hence of solidified things, some are solidified by the absence of the moist, others are solidified [by the absence] of the hot.

That thing the loss of which solidifies something dissolves the same thing by its presence. Hence salt and nitron are dissolved by moist, for whatever is solidified by the absence of 20 the moist, these things the moist dissolves and melts, unless when they are solidified, they bond together and contract in such a way that they do not admit water through their pores because the pores are narrower than the bulk of the water (*tôn tou hudatos onkôn*).[201] Hence, as Aristotle said a short while ago, pottery is not melted by the moist. It ought to be dissolved by water since it is solidified by fire. But [it] is not dissolved by water because of its density (*puknotêta*). Its gradual desicca- 25 tion by the hot brings about the secretion of the water as it changes into air. It is not possible for water to pass through the pores through which the rarer air escaped.

All things which have not been solidified by the hot in the way pottery is, are dissolved by the moist, as nitron, salt, and earth derived from clay (*pêlou*), but not [earth derived] from
30   pottery. Ice and whatever else is solidified by the lack of hot (for example, metals) are both melted and dissolved by hot. Aristotle has said which bodies can be solidified and dissolved and also which can be solidified but not dissolved. Such are pottery and certain stones.

Aristotle next says which things cannot be solidified. He says
35   that the unsolidifiable is 'whatever does not contain in itself a watery moisture nor is of water',[202] but has more earth and hot in it, such as honey and must. These have a lot of the hot in
215,1   them, and are 'as it were, seething' (*hôsper zeonta*). Hence they are thickened by the hot but not actually solidified. Those things also cannot be solidified, he says, which do have water, but have more air. For this reason neither olive oil nor quicksilver (perhaps he means mercury by this) are solidified. Nor
5   do sticky things solidify. Hence pitch and birdlime do not solidify.

CHAPTER 9

Aristotle next says which of the solidified things can be softened. He says that these things can be softened: whatsoever has been solidified, which is not [composed] out of water, but predominantly [composed] of earth. For ice, being [composed] out of water and having been solidified, cannot be softened. But that which is predominantly [composed] of earth and is solidified so that (1) all of the moist has not been evaporated out of
10   it, (as in the case of nitron and salt, for all the moist of these has been evaporated),[203] and (2) it is not irregular (*anômalôs*), as with pottery (he may mean that pottery is irregular with respect to moisture because it has some moisture, but irregularly, or else he means that pottery has irregular pores, because [the pores are] not on straight lines). Whatever, then, is [composed] more of earth and is so solidified that the moist is not
15   completely evaporated from it, and is not irregular in the

arrangement of its pores, but is tractile (*helkta*), so that when stretched it grows longer without having been wetted (such as sinew and leather), such things can be softened. But also, all solidified things which are not [predominantly composed] of water, but which are malleable (*elata*),[204] these too can be softened, but softened by fire, like iron (for it is softened when heated) and horn.[205] Certain woods too, when heated, are naturally prone to be bent (*kamptesthai*), and to be softened.[206]   20 (One might ask whether anything which is solidified out of water is malleable.)

He says that of those things that are melted and dissolved[207] and of those that are not melted, some are capable of being drenched (*tenkta*)[208] and some cannot be drenched; whatever accepts water into itself when it is wetted so that it becomes moister and softer is capable of being drenched, whatever is not affected in this way is incapable of being drenched. Wool and earth are capable of being drenched but cannot be melted, while copper, which can be melted, is not capable of being drenched. (But melted copper is not melted by water, but by   25 fire.)[209]

Aristotle says, 'Of the things melted by water some are not capable of being drenched'[210] (for example nitron and salts). For, when they are wetted they do not become softer while remaining in the same state, as is characteristic of that which is capable of being drenched. He says that generally those things are capable of being drenched which are 'of earth and have pores both greater than the bulk of the water, and harder [than the water]':[211] greater, so as to receive the water through   30 them, and harder, so as not to be dissolved by [the water]. 'Those things are melted by water',[212] he says, which receive the water through the whole of them, and not only through pores, which survive and remain after having taken in the water.[213] He tells the cause of the fact that earth is both drenched and melted by water, while nitron is melted but not drenched. [The cause is] that nitron has pores throughout it,   216,1 and these pores interconnect with (*suntetrêmenoi*) each other everywhere; so that as soon as the water has been taken in the parts [of the nitron] are separated from one another by it. But in earth some of the pores are [arranged] so as to be throughout it, while others lie so as not to meet (*parallax*) and do not

5   interconnect with each other. When, then, the water is received
by the pores which interconnect with one another and which
are throughout [the earth], [the earth] is melted. And when the
water is received by the pores which lie so as not to meet, [the
earth] is drenched but not melted. One might also say as a
consequence of what has been said, that the earth with harder
pores is drenched but not melted, while the earth with softer

10   [pores] is melted. But since the [pores] in nitron are always
softer, they are dissolved upon receiving the water.[214]

Aristotle next states which bodies are flexible and can be
straightened, which are not flexible, and generally what the
capacity to be bent or straightened is, and what it is to be bent
and to be straightened, and that the convex and the concave
are bent.[215] He also speaks of things which can be broken and

15   [things] which can be shattered (*kataktôn kai thraustôn*) and
says that some things admit of one or the other of these, others
admit of both so that they are both broken and shattered. He
explains what each of these is, what the difference between
them is, and what the cause of each of these affections is.

He speaks of impressible and non-impressible bodies, what
they are and why they are [either impressible or non-impres-

20   sible], and what it is to be impressed and not to be impressed.
He says that of impressible bodies, some can also be softened,
like wax, others are harder, like copper.[216] Of the non-impressible things, some are hard, like pottery (for pottery is non-
impressible), some are moist, like water. For water is not
impressed when it yields, but is divided and moves aside
(*antimethistatai*). Aristotle says which impressible bodies are

25   plastic (*plasta*), that is, which submit to moulding (*plasei*), and
which are not plastic, but are compressible (*piesta*).[217] [He also
says] what the compressible is, which bodies can be compressed, and why they are compressible.

He says what pressure (*ôsis*) and impact (*plêgê*) are, and how
they differ. He also says which bodies are tractile (*helkta*) and
what tractility is. But some bodies are both tractile and compressible, while others have only one of these [affections]. He

30   discusses the malleable (*elatôn*) and non-malleable, and says
what they are, and that the malleable are all also impressible
but that the impressible are not all malleable. Of compressible
bodies some are malleable and some are not. Aristotle discusses

the fissile (*skhistôn*) and non-fissile. [He says] which bodies are
non-fissile, and what splitting is and how it differs from cut- 217,1
ting. He also says what can be cut, and that some things can
be both split and cut. For the most part [these can be] split
lengthwise and cut crosswise.

He speaks of the other affections which he enumerated a
little while ago in terms of possibility and impossibility. A thing
is sticky (*gliskhron*), he says, if, 'when it is tractile or moist, it 5
is [also] soft'.[218] Bodies are like this which are so constructed
that their parts interlock (*epallassein*) with one another and
one part is joined together with another as in a chain, and the
parts are not easily separated. Things such as this are tractile,
because they are able to stretch to a great length without
breaking and to contract (*sunienai*) again. Bodies which are not
like this are stiff (*psathura*)[219] and easily shattered.

Aristotle also says which things can be squashed,[220] which 10
cannot be stamped, which are combustible and which are
incombustible and why the combustibles are such. Bodies are
combustible because they have pores which take in fire and
because the moisture in their longitudinal (*kat' euthuôrian*)
pores is weaker than the fire.[221]

After this Aristotle discusses fumable and non-fumable bodies. He says that fumable bodies are those that have moisture, 15
but not the kind of moisture that is evaporated by itself apart
from the body which has the moisture when these bodies are
burned and become heated. Having said that the moisture in
fumable bodies does not evaporate by itself when they are
burned, he discussed vapours, [saying] what a vapour is: an
ejection which is capable of moistening [produced] out of moist
into[222] air and pneuma resulting from hot which has the power
to burn.[223] It is not by fire that the moistness of fumable things 20
is ejected into air when they are heated in such a way that the
things themselves remain (they are indeed consumed in the
process of fuming); it is *time* that does this, for, as they age, so
the moisture is evaporated and they either become dry or
change completely into earth.[224] The ejection of the moist from
fumable bodies which occurs because of time but not because
of fire differs from that which occurs because of fire in that the 25
former is not capable of moistening, nor does it become
pneuma.[225] Aristotle has defined pneuma as 'a continuous

longitudinal flow of air'[226] which, if it received additionally a flow from a certain kind of source, would be wind (*anemos*).[227] When fumable bodies are burned the moist is not ejected by
30   itself; rather, the ejection is produced from both moist and dry air together. Hence this ejection does not moisten (*hugrainei*) [its surroundings], but rather colours [them], for the surroundings are stained by this kind of ejection.

Aristotle says also that the fumes of woody bodies (including bones, hair and leaves) are called smoke (*kapnon*).[228] 'The
218,1   fumes of fatty bodies are sooty (*lignus*), the fumes of oily bodies (*liparou*) are smelly steam (*knissa*).'[229] This is why, he says, olive oil does not boil or thicken, for because it is an oily thing it gives off fumes, but is not vaporised.[230] The moist which is ejected by itself is characteristic of vaporisable bodies, while fumable bodies, when they become fumes, are consumed along with the moist in them.

5   'Water', he says, 'does not fume but vaporises',[231] because the ejection produced from it when it is heated, is moist, but without dryness. He says also that of wines, musty wine fumes, for it is fatty and acts something like olive oil. For it is not solidified by cold, burns just like olive oil, and is not properly
10   wine except in name. It does not behave like wine. For because it does not have a winy flavour,[232] it does not intoxicate, whereas any ordinary wine does. The fumes from it are slight, he says, that is, they are not poured out to any great distance. Hence it becomes flame (*phloga*) as it fumes, because the exhalation becomes fire and [comes to be] flaming on account of its thickness (*pakhutêta*).[233]

15   Aristotle also says which bodies are combustible and which are incombustible, and further, which combustible bodies also produce flames and which do not although they burn. Some inflammable bodies also produce charcoal (*anthrakas*), he says. He states that anything which is fumable but not moist produces flames when it is burned. Olive oil, pitch, and other bodies are moist, hence are not inflammable of themselves but
20   [only] when mixed with and poured on other things.[234] Whatever gives off smoke (*kapnon*) is most inflammable, for flame is burning smoke.

Charcoal-yielding, he says, are those flame-producing things which have more earth than smoke [in them].[235] He discusses the reason (*aitian*) why some bodies that can be melted are not inflammable, while others are inflammable but cannot be melted, and yet others are both inflammable and can be melted. He says which fumable bodies produce flames, which do not, 25 and the cause of this. For fumable bodies which cannot be melted are inflammable, and those bodies cannot be melted which consist mainly of earth. Hence fumable bodies which contain more earth than moist are inflammable, for the earth in them is similar to fire with respect to the dry; therefore whenever it receives the hot, it becomes fire. 'For this reason flame is burning wind (*pneuma*) or smoke',[236] both of which are dry.

The exhalation from wood and woody bodies, as Aristotle has 30 already said, is smoke; the exhalation from wax, frankincense, and pitch (and whatever contains pitch, to the extent to which it does so) is sooty; that of olive oil and of olive-like bodies (that is, greasy things) is a smelly steam (*knissa*). These things burn very little by themselves, he says, because they have in them little of the dry, and the transition and change into fire is due 35 to the dry. For when the dry becomes hot, it catches fire (*ekpuroutai*).[237] 'They burn quickest with something else', he 219,1 says, meaning obviously, among such things, oil and grease, and that the two mix. For the dry, when it has been mixed with that which is greasy, becomes fat (*pion*). (It is clear that by 'with something else' Aristotle meant with something else that is dry.) He says, 'of moist [things], those that are fumable are more moist',[238] while those that ignite (*exaptomena*) and burn 5 are more dry and [so more] earthy.

CHAPTER 10

Having discussed in terms of possibility and impossibility all the opposing pairs of qualities by which he said that homeomerous bodies differ from one another, Aristotle states that homeomerous bodies differ from one another in these affections and these tangible differentia (*haptais diaphorais*). (For all the opposing pairs of qualities mentioned are tangible.) In addition to these [affections] he mentioned [that homeomerous bodies 10 differ] 'in flavours, odours, and colours'.[239] He did not, however, add that some of these [bodies] are noisy and others are noiseless, which hearing distinguishes. [He failed to have said] this either because the primary purpose of his discussion is not to explain the difference between objects of the senses, or because it is manifest that the faculty for perceiving sounds is also a criterion. He had spoken of these matters a little previously.[240]

15   Aristotle enumerates the homeomerous bodies: they are, for example, the metals and whatever is produced from (*ek*) them by separation (*ekkrisei*). For some things are produced from lead, others from quarried things besides. Further, homeomerous bodies are in animals and plants, as that out of which the composition of the anomeomerous (*anomoiomerôn*) [bodies is produced]. And the fibres in plants are classed with the homeomerous [bodies].[241]

20   Now anomeomerous [bodies], he says, have causes other [than those of the homeomerous bodies] (for the proximate matter [*hulê prosekhês*] in them is the homeomerous [bodies], and the agent is the seed). But as regards the things from which they come to be (they are composed, as has been said, out of homeomerous [bodies]), the matter of these [i.e. the homeomerous bodies] is, as has been said previously, the dry and the moist, of which the one is water, the other earth[242] (for water most manifests the moist power, earth the dry power[243]). Their

25 efficient [cause] is the hot and cold (for these bring about the composition of homeomerous [bodies] out of those, that is out of earth and water). Let us therefore consider, he says, among the homeomerous [bodies] which are composed from earth and water as matter, which are preponderantly of earth and which are [preponderantly] of water. [Let us consider], that is, those things which we have already mentioned and which he now goes over.

30   Of the homeomerous [bodies] (whether simple or mixed)[244] fabricated by the hot and the cold, 'some are moist', he says, 'some are soft, and some are hard'.[245] Which [of these] are hard or soft because they have been solidified has been said. Of the moist [bodies] (in which he groups the soft [bodies]) those that evaporate are of water, while those that do not evaporate are

35 either 'of earth or are of both earth and water'.[246] By 'of earth' he means not of earth alone (how could a thing out of earth alone possibly be moist?) but either *more* of earth or equally of

both earth and water. Milk is an instance: although moist, it   220,1
does not evaporate in such a way that it totally expends itself
into vapours, but rather it is condensed.[247] 'Or of earth and
air'[248] as is the case with wood.[249] Aristotle may be speaking of
green wood, for that is also moist. 'Or of water and air',[250] as is
the case with olive oil. Moreover, things that are thickened by
hot are [composed] of both earth and water.

He is unsure whether it is necessary to place wine in the class   5
of moist bodies, or in the class of mixed bodies. In so far as it
evaporates, and such bodies [i.e. those that evaporate] are
moist, wine too should be one of them. But, again, in so far as
wine is thickened by the hot, as is new wine (for when it is boiled
it is condensed, and such bodies are [composed] of both [earth
and water]), it seems again to be out of earth and water. He
solves the problem by means of a division of wines. Those that   10
are thickened by cold are clearly mainly of earth, he says. Those
that are thickened by both hot and cold are compounds (*koina*)
of several things, he says. For olive oil is of water and air, honey
of earth and water. So it is with sweet wine.

Of composed [bodies], that is, those that have been solidified
and are hard, those solidified by cold are [composed] of water,
those solidified by hot are [composed] of earth. Those that seem   15
to be solidified by both hot and cold (whatever is solidified by
being first heated and then cooled is like this) appear to be
solidified by being deprived (*sterêsei*) of moist and hot. For
those that are solidified by fire are solidified by being deprived
of moist, those [solidified] by the cold [are solidified by] being
deprived of hot, and those [solidified] by both [are solidified by]
being deprived of both.[251]

'Of those from which all [the moist] has evaporated ....'[252] Of
the homeomerous [bodies], he says pottery and amber are   20
examples of those from which all the moist has been evaporated
and which have 'received solidity'. But pottery has been solidified by both activities (*poiêtikôn*) while amber, like the other
'tears' of plants (gum, frankincense and myrrh) is solidified by
cooling alone. The river Eridanus,[253] by cooling itself, causes
the hot of the 'tears' to be squeezed out. The hot, when it is   25
squeezed out, brings the moist in 'the tears' [that is, the amber]
with it. Thus [the amber] is solidified by being dried, in the
manner of boiled honey which remains intact (*hupomenei*)

when it is thrown into water. He says: the hot of the honey is ejected, along with the moist, by the coldness of the water.[254] By amber (*êlektron*)[255] he does not mean the mixture of gold and silver but a certain stone which occurs in the river Eri-

30   danus[256] and which, when heated, draws chaff to itself.

All things which are solidified by both [hot and cold] and from which all the moist evaporates, such as pottery and amber, are, he says, solidified primarily by both [hot and cold], such as pottery and amber. These are solidified primarily (*mallon*) by the cold, he says, and they are composed of earth. Of such things

221,1 some, he says, cannot be melted or softened, such as amber and certain stones, like the stalactites in caves. For these are solidified primarily (*mallon*) by cold, because the hot in them escapes (as happens with amber), and because of its departure, [the hot] is a co-producer (*suntelountos*) of the solidification.[257]

5   In the other things solidified by both hot and cold, or solidified by hot [alone], the co-producer of their solidification is the application of fire from without. Those things from which not all of the moist is evaporated are composed mainly of earth; they can, however, be softened. And whatever is melted by fire is more watery, whatever is [melted] by water is [composed] of earth. Whatever is not melted by either is [composed] of earth or of both.

10   Since, then, he says all homeomerous [bodies] are either moist or solidified, and since they [i.e. the moist and the solid] are among the affections that have been mentioned (for of the moist things, some are evaporated, some are not; some are thickened by hot, some by cold, some by both; of solidified and hard things, some are solidified by cold, others by hot, others

15   by both, and some are melted by fire, and some by water) and [since] there is no other affection apart from these in the bodies that have been mentioned, and he has explained, with regard to each of these affections, to what [body] it happens, we have all we need in order to be able to know whether any homeomer-ous [body] is composed of water, or of earth, or of both together,

20   and whether the cold, the hot, or both is the active cause of this [homeomerous body] and of the composition [of it].[258]

CHAPTER 11

Next, considering the homeomerous [bodies], and distinguishing them according to the affections that have been mentioned,
Aristotle shows which are [composed] mainly of water, which
are [composed mainly] of earth and which seem to have a share
of both equally. He has stated that 'it is necessary to pursue
the question what sort of solidified or moist bodies are hot or
cold on the basis of what has been said'.[259] He says about these
generally that whichever of them is watery, is for the most part   25
cold, unless it has foreign heat from without, as do lye, urine
and wine. Lye is hot because it has been strained through
ashes, which have heat, but urine is hot because it has been
produced by the concoction and digestion of food, which is
brought about by heat. And wine is hot because of ripening, and
from the heat of the sun.   30

Again, those [homeomerous bodies composed mainly] of
earth are hot (again for the most part) because they are
produced and fabricated (*dêmiourgeisthai*) by hot, as, for instance, are quicklime and ash. For both are produced by fire,
the one from stones, the other from burnt wood.   222,1

Having stated that those homeomerous [bodies composed]
mainly of earth are hot because of the craftsmanship of the hot,
he then gives the cause of this: for the matter and the substrate
of all such bodies is cold, i.e. [it is] some coldness. For the moist
and the dry are the matter, since of the four powers, these [the 5
moist and the dry] are the passive ones. Water and earth are
the bodies especially associated with these powers, and both of
these are given form by the cold. For water is moist and cold,
and earth is dry and cold. And so all bodies that are of water
or are of earth are cold with respect to their substrate unless
they acquire a foreign heat from without as in boiling water or   10
that strained through ashes like lye.

After saying that ashes have heat, he remarks generally that
there is more or less heat in all things that have been produced
by heat and burning (*purôseôs*). Hence living creatures are
actually generated in putrefying things because that which
causes the putrefying in them is some heat. For the cause of

putrefaction, as has been stated in the first part [of this
15   book],[260] is foreign and external heat, when, because there is so
much of it, it destroys the proper heat in the putrefying thing.

Whatever is [composed] out of both earth and water, he says,
has heat because it has been produced by heat. For most of
these have composition and solidity because of concocting
heat.[261] Some of them are putrefactions, as are liquefied hu-
20   mours (*suntêgmata*). Each such body, [composed] out of both
earth and water, is hot so long as it retains its own nature, (for
this [i.e. heat] is what holds it together and makes it such as it
is): [such is] blood, semen, marrow, fig-juice, and all such things
– but when they are destroyed, and depart[262] from their nature,
and lose their form, they are not hot. There is left in them the
matter, which is earth and water, both cold. Hence such bodies
25   seem to some to be hot, to others to be cold, because whenever
they are 'in their natural state', they are hot, and when they
depart from it, they are cold and solidified. Thus it is, he says,
as has been said, that in all these cases the substratum is cold.
Yet, 'as was determined and said before',[263] those in which the
matter is water are cold for the most part (for water is most
30   opposed to fire, since both its powers, moisture and coldness,
are contrary to fire, which is dry and hot), those whose matter
is of earth or of air are hot.[264]

It sometimes happens, he says, that the same things come
to be both very cold and very hot by acquiring a foreign heat.
For whatever is composed out of water and has been solidified
35   by cold and is extremely solid (*stereôtata*), is especially cold
(because it has been solidified by being deprived of heat), and
223,1   burns (*kaiei*) severely whenever set on fire. For thicker things
burn more when exposed to fire; water [burns] more than
smoke (boiling water burns more than flame, which is burning
smoke) and again stones more than water.[265]

CHAPTER 12

Aristotle has said what the substrate is in homeomerous bod-
5   ies, and what each of the homeomerous bodies is mainly [com-
posed of]. Since flesh, bone, sinew, and such things are among
the homeomerous [bodies] (for out of these the anomeomerous

parts of the animal [are composed], out of which the animal [is composed]), he proposes next to say with respect to these what each is, not according to the substrate (*hupokeimenon*) (for he has done that), but according to the definition (*logon*).[266]

For we now have, he says, the classes (*genê*) of homeomerous [bodies] according to what each of them [is composed] of (earth,   10
water, or both), since we grasped [these classes] from the generation of [homeomerous bodies]. For the substrate from which they [the homeomerous bodies] are generated is the substrate which they have when they have been generated. The nature of the anomeomerous [bodies] is produced out of these, for the anomeomerous [bodies] are [composed] out of the homeomerous [bodies]. For first the homeomerous [bodies] are [composed] out of the elements and then out of these as substrate and matter [come] 'the complete works of nature',[267] namely, animals and plants. 'Out of these as matter',[268] he said,   15
for the form (*eidos*) of them [i.e. the anomeomerous bodies], with respect to which most of all they are what they are, is not such [i.e. is not matter].[269] He says that all homeomerous [bodies] are composed out of water and earth (which were mentioned before) as matter, but as for their substance (*ousian*), that is (*kai*),[270] the being which each of them is by definition, this is in the definition and is what each of these is in form.

Aristotle says that it is clearer, in the case of the latter,   20
meaning the anomeomerous [bodies] (for they come after the homeomerous [bodies] if indeed[271] they are compounded [*sunkeitai*] out of them), that their being is [determined] by their definition, and completely [clear], he says, with respect to those [anomeomerous bodies] which are like instruments (*organa*) of the body and seem to have been given to us for the sake of some purpose. It is obvious with these that their being is in the definition. For their being is in [their] capacity (*dunasthai*).

He showed that it is more clear [than it is in the case of   25
homeomerous bodies] that the being of anomeomerous [bodies] is [determined] according to their definition, taking the dead person [as an example]. For it is more obvious that a dead person is called a person equivocally, and not in the strict sense. This [i.e. a person] is that to which the definition applies, but

the definition of person does not apply to a corpse. For no one
30   defines (*horizetai*)²⁷² a dead person as a rational mortal animal.
In this case it is more obvious than with blood and each of the
[other] homeomerous [bodies]. For in the case of these, too, the
things said of the living are used equivocally of the dead, only
it is not apparent in the same way in the case of [the
224,1   anomeomerous bodies]. For just as a dead person [is called] a
person equivocally, in the same way a hand of a dead person
[is called] a hand equivocally.

Aristotle began with the last [of the things produced by
nature] (this was the complete animal), and then returns to
those [produced] before [the animal], of which the anomeomerous [bodies] are next to complete [animals], and before these,
5   the homeomerous [bodies]. He showed from the example of
stone pipes²⁷³ that things which exist for a certain purpose,
when they no longer are able to fulfil that purpose (*poiein*) and
retain only their shape, are spoken of equivocally.²⁷⁴ Just as
pipes are instruments, so too the hands and all the parts of the
animal are instruments of a sort which have come to be by
nature for the sake of a certain purpose. It is less clear, he says,
in the case of flesh, bone, and [other] homeomerous [bodies],
10   than in the case of the hand, the face, the foot, and [other]
anomeomerous [bodies], that they are instruments, and less
clear to what use they can be ascribed. It is even less clear in
the case of simple bodies, which we call the elements, such as
fire, water, and the rest. For it is least of all clear what the
particular purpose is of each of these bodies, in which matter
plays the greatest part, that is, which are nearest to matter.
15   Those are the nearest to matter which are the first to be
produced out of matter, and these are the elements. Hence in
compounds these have the definition of matter.²⁷⁵

Therefore, if the extremes out of which the complex [is
produced] (these are the matter and the form) should be taken
separately, matter is nothing other than itself (for it is the
ultimate [*eskhatê*] matter, the first substrate of bodies; for such
20   a nature is simple), so also the form, i.e. the substance (Aristotle calls form 'substance', since the being of every compound
is according to [*kata*] form), [is nothing other than itself]. And
form is nothing other than the definition (*logon*) according to
which matter is given form. The intermediate forms [derived]

from these [two] are more like that to which they are nearer, those nearer matter being more material, those nearer substance and definition being more formal.

Those things are nearer to definition and form which are   25
compounded (*sunkeitai*) out of bodies which have already been given form, in so far as these are given form. For example, the face is compounded out of certain homeomerous [bodies], in so far as these have been given form (for [it is compounded] out of eyes, in so far as they are eyes, and so with the other [parts]).[276] Those things are [nearer] to matter which have underlying matter separate from forms.[277] For such is the [material] substrate (*hupokeimenon*) of the elements, since each of the simple   30
primary bodies, he says, is for the sake of something (*estin heneka tou*), and since each of these [the four elements] is not [merely] water or earth taken in no matter what condition, nor something [else, that is, neither air nor fire], just as neither flesh nor liver [is flesh or liver taken in no matter what condition]. But just as [flesh and liver are flesh and liver] when they produce that which is expected from them (*hotan to hautôn parekhêtai*), so too [the elements are elements], when each of them retains its own nature and its particular function (*ergon*).[278] 'Of these even more',[279] he says, meaning more than flesh and each of the internal organs and generally the ho-   225,1
meomerous [bodies], the 'that for sake of which' is more evident in the case of the face, the hand, and each of the anomeomerous [bodies].

Aristotle has stated that each of the anomeomerous [bodies], like the homeomerous [bodies], but even more so, has 'a that for the sake of which', and that the substance of these [anomeomerous bodies] is defined/determined (*horizesthai*) according to this. He now says what this is, 'for the sake of   5
(*kharin*) which' each of these [anomeomerous bodies] has come to be, is, and is given form. All such things, he says, are defined/determined (*horizesthai*) by some function (*ergôi*). Those that can perform the function proper to them are truly that which they are called (for an eye is one that performs its function or can do so, and the function of an eye is seeing), while those that cannot perform the particular function of the thing the name of which they have, to each of these the name is   10
applied equivocally, as it is to the eye of the dead person, or the

stone [eye]. Nor would a wooden saw be [so named] correctly, but only equivocally, just as the likeness (*eikôn*) of Socrates [is called] Socrates [but equivocally]. For the wooden saw is a likeness of a saw. As it is with these things, so is it of course with flesh and blood and similar parts. But because the func-
15 tion of these is equally not familiar in the same way, it is unclear which of them are properly what they are called, and which are called [flesh, blood, etc.] equivocally after the others.

It is even more unclear in the case of simple bodies what their natural function is and by what they are defined/determined (*horizetai*).[280] So it is with the parts of plants, he says, for they have some proper function by means of which the being (*to einai*) of them is defined/determined (*horizetai*). But among
20 natural and homeomerous bodies the ones without soul are also defined/determined (*hôristai*) by some function, such as copper, silver, gold and each of the others. All natural bodies are defined/determined (*horizetai*) by some power, either active or passive. For fire is defined/determined by doing something, earth by undergoing [something], as are flesh, sinew and the parts of animals in general.[281] But the definitions (*logoi*) of them, that is the powers of them, according to which their
25 substance (*ousia*) is defined/determined (*horizetai*), are not known very accurately. Hence it is not easy to judge and to distinguish when this particular power belongs to them, [so that] they are properly what they are called, and when they are only so named equivocally. [This is true] unless the power by which each of them is what it is called should be completely lost, so that only the shapes or the colours of the bodies remain,
30 and these but superficially, in the same way as the bodies of the long dead seem to preserve their shape for a certain time and then suddenly to become ashes. So too ageing fruits are such in shape alone and no longer move the senses as fruit do. This is true also for the things solidified out of milk.
226,1 Having said these things, and having shown that the substance (*ousia*) of homeomerous bodies (of the parts of animals and of other [homeomerous bodies]) is defined/determined (*horizetai*) according to form and definition (*logon*), Aristotle states what the definition (*logos*) of the homeomerous parts of animals is, and by what they are given form. He says that they are
5 defined/determined by heat and coldness and by the motions

from these.[282] For each of these [animal parts] acquires both its solidification and its composition (*sustasin*) by means of the hot and the cold.

He says what the motions from the hot and the cold are: 'tensility (*tasei*),[283] tractility, fragility (*thrausei*),[284] hardness, softness and the rest'[285] about which he spoke in defining/ determining (*diorizôn*) what the passive powers and their [corresponding] incapacities are. For the differences these 10 [homeomerous parts of animals exhibit] with respect to one another are produced according to these [powers and incapacities]. For some [of these differences] are brought about by hot and the motion from the hot, some by cold, and some by both.

Those bodies which are composed out of these (meaning [out of] the homeomerous [bodies]), which are the anomeomerous [bodies], could not be said to differ from one another through these [powers and incapacities].[286] For the head could not be said to differ from the hand or the chest from the foot in tensility, tractility, fragility, hardness, softness or any such 15 thing, although they have their being through some one of these. But just as the cause of the generation of gold, silver, iron and each such thing is coldness or heat, and the motion from one of these, but none of those is the cause of the coming into being and the being of the things which are [composed] out of these, such as a bowl or a saw, so is it with the anomeomerous 20 [parts] of the animal, which are composed out of the homeomerous [parts]. But there is a difference, because the cause of the bowl and the saw and the things made out of iron and silver is a craft (*tekhnê*), while the cause of these anomeomerous [parts] is nature or some other cause. For a person begets a person by nature, and a horse [begets] a horse [by nature].

Having said these things, Aristotle returns to the discussion of the homeomerous parts. For he has proposed to speak of 25 these here, saying what each is, on the one hand, with respect to matter, and on the other, with respect to definition (*logon*). He says that we grasp from what has been said to which class (*genous*) each of these [homeomerous parts] belongs; for they are either some moist or dry or [composed] out of both.

For he has discussed and shown what sort of things are moist or dry or mixed out of both, and how they are composed. It is necessary, he says, when considering each of such [i.e. the 30 homeomerous] parts to give an account of what it is (*to ti estin*), taking its class (*genos*) first: for example, to say that moist or dry [is the genus], adding then what sort of thing it is and how it is generated. This is how we know what each thing is and how we give an account (*logon*) of it, considering it either according to the substrate, that is (*kai*) matter, or according to the definition (*logon*), that is (*kai*) the form, especially when we know both.[287]

227,1

These things are in agreement with Aristotle's statements in the first book of the *On the Soul*. For there too he said that some people give definitions from the matter and some from the form, but that those definitions of natural things are most complete (*teleiotatous*) which give the form of the thing in question and add that it is in such and such a matter.[288]

5

Having said, 'For in this way, we know why each thing is, and what it is',[289] he says that it is necessary also to know the causes of both generation and destruction, the final (*telikên*) [cause], which he called 'why it is', and the material (*hulikên*) [cause], which he indicated with the words 'and what it is'.[290] He added that it is necessary to know the efficient cause (*poiêtikon*) with the words, 'and the source (*pothen hê arkhê*) of motion'.[291] Hence it is possible to understand everything from what has been said by asserting that each of them is either moist, dry, or both at once, [and so that] the matter is thus [moist, dry or both], and that the form is whatever kind of solidification or composition (*sustasin*) it is. This can be grasped from what has been said by discovering what [kind of solidification or structure] 'fits' them. The efficient cause [can be grasped by saying] that such a thing comes to be, or has come to be, by hot or cold or both.

10

15

Since these homeomerous [parts] have become better understood in this way, it is necessary, he says, to speak next of the anomeomerous parts in a similar fashion, and then in the same way of the things which are composed out of them, which are plants and animals. The *Parts of Animals* seems to follow this book. For in the second book of the *Parts of Animals*, Aristotle spoke of the matters which he here says require discussion. For he speaks there first of the homeomerous parts, and then of the anomeomerous [parts], which are composed out of the homeomerous parts.

20

Notes

1. See Introduction for a discussion of the question of the authenticity of *Meteor*. 4.

2. Here, and throughout this work, Alexander calls the four most basic contraries powers (*dunameis*), and not qualities (*poia*), the term usually found in Aristotle. The four contraries consist of the pairs of contrarieties, heat – coldness (or the hot – the cold) and dryness – moisture (or the dry – the moist). I have endeavoured, whenever possible, to mark the differences found in the Greek between the necessarily substantive and adjectival forms of these terms (the difference between, say, 'heat' and '(the) hot'). It is these four powers which the four basic elements (earth, air, water, fire) are 'out of', and so, consequently, is everything else. See Introduction for more on the 'out of' locution, and a full discussion of the relationship between the elements and contrarieties. See *GC* 329b7 for a similar use of 'tangible' (*haptos*), and *GC* 330a31 for a similar use of 'coupling' (*sunduazô*).

3. Earth, water, air and fire.

4. See *GC* 2.1-3.

5. Alexander uses the phrases 'the first bodies' (*ta prôta sômata*) and 'the elements' (*ta stoikheia*) as synonyms. See also *GC* 329a29, 330b6.

6. This word (*eidopoieô*) is not found in Aristotle. The closest we find is at *GC* 329b9 where, in a discussion of what the basic contrarieties are, Aristotle claims that not all the contrarieties 'make forms and principles' (*eidê kai arkhas poiousin*). This term is found in other works of Alexander (e.g. *in Metaph*. 1, 57,7; 54,8; 59,15, etc.), where it means, as in the rest of this text, 'to give or confer [a] form'.

7. *GC* 329b17.

8. Namely the combinations heat-coldness and dryness-moisture, since these are pairs of opposites.

9. Here all four contrarieties, and not just the active two, heat and coldness, are said to confer form. See Introduction for a discussion of activity, passivity, and the contrarieties.

10. Here Alexander follows closely *GC* 2.3, 330a30-b5, with minor changes which follow the heavily doctored manuscripts EL, as against the preferred FHJ.

11. On the importance and development of this view, see Introduction.

12. 329b24f.

13. 378b15-16. For a discussion of the meaning of 'determine' (*horizô*) see Introduction.

14. I use 'undetermined' as opposed to 'indeterminate' to emphasise the fact that these things have not undergone a process of determining. In addition, 'indeterminate' has misleading metaphysical connotations.

15. Alexander uses *idios* (one's own) and *oikeios* (proper) as synonyms.

16. *henousi* is from *henoô*, a rare word. Best sense is made of the assorted things that the hot and the cold do by taking the *kai* to be epexegetic. By doing so one can also understand why we here find *henousi*, for it explains the rather opaque *sumphuô* found in Aristotle.

17. The discussion here concerns homogeneous things, as opposed to things that are homeomerous, which will be discussed at length later. Homogeneous things are things that are alike in kind, regardless of their internal composition. In other words, 'homogeneous' is a comparative term. Two things are homogeneous if they are the same in kind, both being, say, iron, or marble. Homeomers are things whose parts are the same as the whole, things that are compositionally homogeneous. All the parts of flesh, or iron are flesh and iron respectively, and so both are homeomers. Many homogeneous things may be homeomers, but they need not be. Two apple pies are homogeneous, since they are both apple pies, but neither are homeomers. A piece of iron and a piece of copper are each individually homeomers, but together they are not homogeneous.

18. See *GC* 329b25f. and 336a1.

19. Here I follow Coutant and Hayduck in excising in 181,1, '*pegnutai de to hugron*'.

20. It is unclear what, if any, circumstances have it that the hot and the cold are by themselves, since all actual bodies are out of all four powers, as Alexander realises (see 182,1 below).

21. This is perhaps a pun based on the dual meanings of *horizô*, 'to determine', and 'to define'. Particularly in 4.12, it will be difficult to decide which is the intended meaning.

22. See n. 270 below for an account of the relationship of *ousia* to *to einai*.

23. This follows closely 378b25-8, favouring in line 27 MSS HN against WB.

24. 378b28-9.

25. The discussion here concerns natural destruction, which is just below (181,31-2) contrasted with violent and unnatural destruction.

26. 379b30-1, following MSS HW and Olympiodorus.

27. This is the first use of *hulê* (matter) in this work. For more on the relationship of matter to the passive powers see Introduction.

28. Here again we have talk of passive powers.

29. Note here the strong emphasis on the fact that it is the hot and the cold that act, while the moist and the dry are passive. Even the drying of something is here said to be due to the action of the hot, not to that of the dry.

30. The matter under consideration here is plant or animal matter (leaf or flesh, for instance), and not the passive powers themselves which are matter for the active powers. This explains the use of 'in' here.

31. Following Coutant, I prefer 'simmering' to the often suggested 'parboiling' for '*molunsis*'. See n. 73 below for why.

32. This is the first occurrence in this work of *sunistêmi*. For a discussion of this important term see Introduction.

33. This addition of Coutant's, following Hayduck, has Alexander follow 379a8-9 more closely.

34. See 4.12 below for more on the relationship between natures, moving principles, and the hot.

35. 379a17-8, with very minor changes.

36. Even frozen things contain some of the hot, since they are composed of all four elements.

37. This claim is not found in Aristotle, and seems both inconsistent with what is found, and with the facts of the matter. Boiling water does heat the surrounding cooler air, and does not extinguish the lesser heat of the air, but contributes to it. According to Alexander, boiling water could cause putrefaction, by drawing out the (lesser) heat in something near it, but could not itself be putrefied.

38. 379a3-4, with one word-order change.

39. See *Probl.* 966a20.

40. For the importance of this paragraph for the question of the authenticity of *Meteor.* 4, see Introduction.

41. Alexander's point being that the hot usually produces concoction, the cold inconcoction.

42. This is an important claim. We are not to take it that the six processes mentioned literally exhaust the types of concoction and inconcoction. They are but convenient subdivisions. True, the ripening of a fruit is a concoction, and is a ripening, yet not all concoctions which are ripenings are just like the ripening of fruit.

43. 379b19, with very minor changes.

44. See *GC* 321b35ff.

45. It is not the whole organism which is said to be nourished, but the material parts. So unconcocted nutrition, say a just-consumed banana, is unlike the flesh, bone and blood it will become after it is concocted.

46. 379b20.

47. See *Metaph.* 8.4, 1044a15f. on 'proper matter', and its relationship to that which something is composed of.

48. Coutant tells us that: 'The utility of pus, tears, etc. is that they can be easily sloughed off in the form they have attained, whereas otherwise they would harm the body if consumed by it (Cf. *GA* I 18, 725a3-7). Hippocrates (*Aphorisms* I 22, Adams II 199) speaks of the suppurative effect of heat as beneficial (*Aphorisms* V 22, Adams II 238)'.

49. The 'to us' is perhaps surprising, and is not found in Aristotle. Alexander seems to take here a very anthropocentric view of natural teleology, claiming that something is perfected (and so concocted) whenever it becomes more beneficial to us from having been less so.

50. One here expects 'ripened' (*pepainomenon*), not putrefied.

51. 379b32-3.

52. See esp. *Phys.* 2.1, 192b21.

53. *pericarpia* are the skins or casings of fruit. They are said to encase the nutrition in fruits, and so 'within' for '*en*' in 188,22.

54. 380a12.

55. *apoteleô*, particularly in this context, implies the ability to create a perfected wholly formed thing.

56. 380a14-15.

57. 380a20.

58. This paragraph is a commentary on a difficult line in Aristotle, 380a20. Here Aristotle *does* seem to say that matter is determined by natural heat and coldness. Yet Aristotle thinks that cold is responsible for *in*concoction, not concoction. Alexander realises this ('meaning not ...'), and offers a not completely satisfactory explanation, namely that the cold, in so far as it is mastered (as matter?) in concoction, contributes to concoction. (This view of the cold as matter seems to be based on comments Aristotle makes in *Meteor.* 4.5, where he says that the cold is more like the passive powers, because it is an affection of the two passive elements, water and earth.) Also in *Meteor.* 4.5, Aristotle states that cold can heat, or at least contribute to heating, by concentrating heat by surrounding it. Yet Alexander does not here avail himself of this explanation. I prefer a third alternative. Aristotle has just claimed that ripening is, in cases other than fruit, a process like that in fruit, with the term being applied but metaphorically. Then we have the line, 'since, *as we remarked earlier*, there are no specific names for each type of perfection in that which is determined by natural heat and cold' (380a20, my emphasis). What Aristotle remarked earlier (379b12-17) is that the three classes of concoction *and inconcoction* are applied but metaphorically to all cases (this is the gist of the passage). What we have here at 380a20 is a lacuna. What is required is something like, 'since, as we remarked earlier, there are no specific names for each type of perfection in that which is determined by natural heat, and [each type of imperfection in those inconcoctions brought about by] cold ...'. Alternatively, perhaps following MS E, and omitting 'and cold', would help.

59. 380a20-2.

60. Pneuma plays a complex role in Aristotelian chemistry and biology. Here, and throughout this work, it seems to be used in its non-technical meaning as 'air'.

61. 380a23, omitting a *ton*, as in Olympiodorus and 2 MSS.

62. Aristotle defined concoction as a process which makes things thicker, due to the action of heat (380a5f.). Ripening, being a concoction, thereby also has this effect.

63. 380a25-6.

64. That is, either the moist which would ripen were there sufficient natural hot, or that which will ripen when the natural hot increases (fruits start raw, but ripen as the natural hot in them increases).

65. Here we face one of the many difficulties with translating *sunistêmi*. 'Condense' is quite tempting (and is what I have here used), yet it forces one to have it that Aristotle and Alexander think (falsely) that thickening and condensing are equivalent. This then tempts one to translate 189,17 as 'condense, that is, thicken'. Yet this then leaves the obvious problem that many fruits upon ripening, become thinner and softer, not harder and denser. See also n. 70 below.

66. Here Alexander seems to equate thickening with structuring.

67. 380a32-3.

68. Here the comparison is between the liquid and the solid parts of raw things, and so *hai sustaseis* seems here to be best translated by 'pulps'.

69. Not to be confused with excreta (*perittoma*) which are ripe, not raw. It is perhaps of interest to note that excreta are concocted, while urine is not.

70. Here we have a close relationship between things that can be thickened by the hot, and things that can *sunistêmi* due to the hot. Therefore I have translated the preceding four occurrences of cognates of *sunistêmi* by cognates of 'condense', since condensing and thickening seem to be closely related. See also n. 65 above.

71. Raw pottery is unfired pottery.

72. Since both the milk and the heat are within the animal, Alexander construes this as a case of moisture being mastered by internal heat. Since the moisture at issue is the moisture of the milk still in the cow, Alexander's analysis is correct only if the moisture of the milk is considered to be [part of] the moisture of the animal. Alternatively, one may think that this analysis is false from the perspective of the milk in the cow, and so rests on an equivocation on 'within'.

73. Following Coutant, I prefer 'simmering', to the often suggested 'par-boiling', since par-boiling is boiling, just briefly (as is scalding), while simmering is an imperfect boiling (and so not a boiling at all), where the heat is inadequate to bring about true full boiling. The difference between an inconcoction and its associated concoction is not that the inconcoction is a brief concoction, but it is different in kind. Actually this term does not correspond neatly with any in English. It is like simmering when the heat in the surrounding moist is insufficient, yet if the moist in the thing is too great to be mastered, even if immersed in fully boiling water, still no concocting takes place. See 195,11f. below for more on this.

74. Alexander is here concerned with the (still) undetermined part of the moist present as matter in things. Something boiled, say flesh or greens, might already have had (in fact, needs to have had) some of the moist in it determined, yet not all of it. Boiling will determine more of the remaining undetermined part of the moist in it.

75. One might expect to read 'things are boiled by (*hupo*) the hot-moist and in (*en*) hot water'. The 'by hot-moist' in the preceding line, and the 'by moist heat' in 191,28 below seem to confirm that here Alexander merely slips.

76. 380b16, itself commenting on 380a29. Here Alexander, following Aristotle, is giving an elemental description of what it means for something to contain undetermined moisture. The undetermined moisture must be airy or watery, since both air and water are characterised by being moist. Alexander makes it clear that he thinks that this element talk is merely substituted for contrariety talk, when he claims, in the next line, that 'fire' here means 'the hot'. See 189,29-34 above for Alexander's commentary on 380a29f.

77. See 192,1 and 196,11-12, where frying is said to be a form of roasting.

78. The density at issue here seems to be that of the moisture in bodies that cannot be boiled, not the bodies that cannot be boiled themselves, since Alexander goes on to claim that it is the moisture that is 'naturally condensed in structure' in things which cannot be boiled. Aristotle's claim at 380b25 that there are bodies with no moist in them (e.g. rocks) is curious. From *GC* we know that all bodies contain all four of the elements, and so contain, in some sense, the moist. The issue here may be whether bodies like rocks are *manifestly* moist. Olympiodorus (291,32ff.) glosses this passage as concerning bodies which have virtually no moisture.

79. This condensed moisture in wood may be sap.

80. Aristotle claims (380b14-15) that only things cooked by boiling are strictly said to boil, since gold, wood, and the like, are not cooked, they do not boil in the strict sense.

81. This is a non-technical use of mixture, since this moisture is foreign, and so not essentially part of the composition of the homeomer (wood, gold, etc.) which does also contain moisture essentially, or properly (yet this moisture cannot be mastered, or else these things would actually boil). The moisture would have to be actually part of the composition of the homeomer for it properly to be said to be mixed with it.

82. The point here is that things like gold or rocks do not contain their own proper moisture, but they may have in them foreign moisture, which can be driven off. This is *like* boiling, but is not boiling, since the moisture driven off is not proper to the body at hand. Note Alexander's reference here to 'things mined', a link with the end of *Meteor*. 3, though this term in not found in the passage in *Meteor*. 4 (380b25f.) that Alexander is here commenting on.

83. Perhaps Alexander has in mind the production of charcoal, which is made from burning wood. While burning the wood ejects its foreign moisture (sap, etc.), resulting in an almost smokeless wood, the charcoal.

84. 381a2.

85. Milk and must are not boiled by being placed in a hot liquid, they are the liquids which become hot when heated by an external fire. However the moist in them is affected as in things boiled, and so they are said to boil. In a sense, water does not boil, it does the boiling.

86. A rather lengthy exegesis of a sentence in Aristotle (381a7f.) thought to be suspect by Thurot and Webster, which seems simply to rely on the fact that boiling is a type of concoction.

87. There are two possible explanations as to why drugs do not, strictly speaking, boil. First is that things which strictly speaking boil are cooked by boiling, and cooking is a process intended to make the cooked things 'useful for eating or drinking or for the benefit of' a nutritional aim. So, if drugs are not for any nutritional end, they are not properly speaking cooked, and so not properly speaking boiled. Second, the boiling of drugs might not even refer to something done *to* them, but something they do, they boil and seethe, akin to comic depictions of witches brews. If so, their boiling would be more like that of milk and must. This is strongly suggested by the active, not passive form of the infinitive found here (and in 193,31.33, *hepsein*).

88. 381a4-5.

89. Following Coutant and MSS B1W in excising *kai tauta* in 194,4.

90. The point seems to be that they become proportionally heavier, that is denser. In other words, things properly boiled either thicken while remaining the same in volume (porridge, for example), or do diminish in volume, yet become proportionally heavier, that is denser.

91. 381a5-6.

92. See n. 90 above for the meaning of 'heavier'.

93. Oil is often absorbed by that which is being cooked in it. The point here is that even if all the oil is not absorbed it does not change its form while boiled, and so the added 'preserved'.

94. The '*te kai*' here, and 2 lines later, is used to indicate that there are not two processes, boiling and concocting, but one, boiling, which is a concoction.

95. 381a12-13.

96. 381a15, itself referring to 379a19.

97. 379a19. See 184,4f. above for Alexander's discussion of this passage.

98. 381a16.

99. Aristotle (381a15-22) does not make the claim that the cold becomes kinetic, only that the heat is 'beaten off' and that this motion differs from that in boiling.

100. 381a17.

101. It is this suggestion that simmering can take place due to the excessive amount of moisture in the thing attempting to be boiled, and *not* due to the insufficiency of the hot in the liquid this thing is immersed in, that demonstrates that neither 'simmering' or 'par-boiling' captures fully the meaning of *molunsis*. When this 'state' is due to excessive moisture in the object, it is akin to par-boiling (a partial, uneven boiling). When it is due to insufficient heat in the surrounding liquid, it is akin to simmering (a 'sub'-boiling).

102. 381a12-13.

103. In such cases one is not boiling something, but roasting it, even if one does immerse it in boiling water.

104. The point being that the inner moisture is trapped by the pores on the surface contracting due to the roasting.

105. See Introduction for the importance of the passage this is commenting on (381b1f.) for deciding the question of the authorship of *Meteor*. 4.

106. This natural boiling is, in fact, digestion.

107. The stomach.

108. Coutant offers the Hippocratics as those who may have thought this.

109. The intestines.

110. Two species of parasitic worms. The rare use here of *terêdon* (lit: 'borer') to name not a woodworm, but an intestinal worm, may suggest that these are hookworms (a kind of nematode) which, like wood worms, bore into their host. Helminthes, according to Peck's commentary on Aristotle's *HA* 5.19 551a8, include trematodes, nematodes and cestodes. I thank R. Sharples for enlightening me concerning such parasites.

111. This passage comprises *Probl.* fr. 241 Rose 3.

112. This account of searing parallels that of simmering.

113. 381b21.

114. 381b22.

115. See Theophrastus, fr. 172 FHSG (= Galen *in Hipp. Aphro.* 14 [vol. 17.2, 404,12-405,3 Kühn]) for a similar sentiment, said to be from his work *On the Hot and the Cold*.

116. I take the gist of the passage from 'the affections in bodies' (198,11) onward to be as follows: Alexander is puzzled as to what is the correct analysis of what it is to be a body qualified in a certain way. His confusion is compounded by *Meteor*. 381b27, which he is commenting on. Here Aristotle says 'and all will be either in actuality, or in the opposite sense'. The question is, all what? The two obvious candidates are affections and bodies. In Alexander's comments he surveys both possibilities, yielding four analyses. He realises that there are two related ways of analysing 'X is a dry body'. This could be explained by the dry's being present in this body actually. On this account 'X becoming a dry body' would be analysed as 'the dry's going from being present in potentiality in X to being present in actuality in X'. This account emphasises the affection and its role, treating the subject or body as a fixed unchanging substrate, and is equivalent to taking Aristotle's 'and all' to refer to affections. Alternatively, one could emphasise the fact that it is a subject, a body, that is dry, or becomes dry, while remaining the same subject through the change. This is rendered by 'all the bodies will be called body [of a certain sort] ... by being in the affections in actuality ...'. This is equivalent to 'and all' referring to bodies. Alexander tempers the oddness of this account (which is just the strict parallel of the first account) by claiming that this is but to say that X is exemplifying the activity of dry, 'that is to say, being like this in activity'. The move to *energeiai* (activity) from *entelekheiai* (actuality) is therefore meant to defuse the possible suggestion on Aristotle's part (i.e. reading 'and all' as referring to bodies) that the bodies actually exist in the affections (a result that Aristotle, in *Cat.* and elsewhere, is at pains to avoid), in favour of an account that has it that the bodies exemplify certain behaviour or activity when characterised by a certain affection. Alternatively, the 'Or: all the bodies ... able [to be like this in activity]' is meant only to be an alternative formulation for stating that a body has a certain affection (as opposed to presenting an alternative way in which bodies may be related to qualities), since, as R. Sharples points out, Alexander is willing to talk of things being in states, but possessing the potentialities for them (see *Quaest.* 2.15 60,12ff.)

117. 381b28-9.

118. Note how here Alexander opts for the account which has it that it is the affections (*pathê*) which are in the bodies actually or potentially.

119. The referent of 'all the others' must be all other affections that are discussed in this work. The claim that all affections are compounded out of the dry and the moist is worked out further in 4.8 below.

120. 381b29.

121. 381b29.

122. *DK* 31 B 34.

123. Perhaps the *te kai* here should be rendered as 'and so', stressing the close connection between becoming composed and becoming determined. See Introduction for more on this.

124. Alexander takes it that composition out of neither has been ruled out.

125. This is an important, yet ambiguous clause, since one must supply the missing verb. Straightforward identity of the moist with water and the dry with earth is not, I think, what is intended. This is a gloss on *Meteor*. 382a2-3, where Aristotle claims that 'Of the elements earth has the proper character (*idiaitata*) of dry, water of moist', and this is what I take Alexander to be saying, albeit in compressed form, so perhaps 'correspond to' is a more neutral translation. Opposed to this suggestion is the *ha estin* in 199,22-3, relating the moist and the dry to water and earth, respectively. In any case, this doctrine in and of itself conflicts with *GC* 331a3-6, where air is said to be properly moist, water cold, earth dry, and fire hot.

126. 382a3. 'Here' means in the sublunary region, as opposed to the heavens.

127. This is a plausible explanation of why Aristotle, at 382a4, states that determined bodies *here* (*entautha*) need be composed of both earth and water. We are told in *GC* (335b32f., and all of 2.8) that all sublunary bodies are necessarily composed of all four elements, while heavenly bodies such as stars (being an example of 'divine things') are composed purely of one element (the fifth element, aether).

128. Precisely what is at issue here is uncertain. The parallel text in Aristotle (382a6-8) concerns where animals are found, or exist. This is what tempts one to read here 'generated *on* the earth and water alone', although the Greek is more naturally rendered by '*in* the earth and water alone'. Yet Alexander's emphasis on generation may indicate that what *he* thinks is at issue is what animals are composed out of, what they come to be in as their material substrate, not the immediate spatial surroundings of where they are generated. This may in fact be Aristotle's intended meaning, referring to matters discussed in *GA* (2.2, etc.). Alexander may be drawing a parallel between what something is composed of, and where it is found, having *en* do double duty.

129. 382a12-13.

130. The use here of 'determinate' and 'indeterminate' is not the technical meaning with regards to informedness and structure or composition, but concerns strictness of definition. Things not absolutely soft or hard are not determinately (= without qualification, *haplôs*) soft or hard.

131. See *DA* 418b26-7.

132. The text reads, 'Hot, hard, soft, hard, moist'. I follow Coutant in emending the first 'hard' to 'cold'.

133. The above account of the absolutely v. relatively hard or soft is problematic. It seems that everything will be absolutely hard or soft – absolutely hard if harder than touch, absolutely soft if softer than touch. What then of the relatively hard and soft? I am tempted to supplement the text as found as follows: 'So that which exceeds the mean which is touch in respect to hardness will be [perceived as] hard without qualification, and will be apprehended by touch as unyielding. Conversely, that is [perceived as] soft without qualification which the sense [touch] lays hold of because it yields to it in a greater degree. Hence that will be [perceived as] hard without qualification which is harder than the touch which lays hold of it, and that will be [perceived as] soft without qualification which is softer than the touch which lays hold of it.' This has Alexander offering an explanation as to why anything harder than touch appears simply to be hard (like, say, both a piece of wood and a rock), while in fact, one is harder than the other. However, there is no manuscript precedent for such an emendation. I thank R. Sharples for drawing my attention to this problem.

134. Alexander's text of 382a21 seems to have read *'lepton'* (delicate) (as with a minority of our manuscripts), as opposed to the preferred *'elleipon'* (deficient) (as with a majority of our manuscripts).

135. One should note the misprint in Fobes (382a23), which reads *eina*, instead of the correct *einai*.

136. This is traditionally rendered as the efficient cause.

137. The preceding is a very close paraphrase of 382b2-3.

138. This schema is problematic. If water is given form by the moist, and earth by the dry, then fire could be given form by the hot. Yet this would leave air (hot-moist) with no contrary to uniquely give it form. Theophrastus modified the Aristotelian schema so as to have fire given form by the hot, air by the cold, water by the moist, and earth by the dry. Such a schema, as R. Sharples has informed me, was also followed by the Sicilian medical school.

139. i.e. 378b21-22, where the hot and the cold are said to be active, and 379a19-22, where cold, along with heat, are said to cause decay, and so be destructive powers.

140. 382b7.

141. 382b7. The sense in which the cold may be an active cause only 'incidentally' is perhaps best explained by Aristotle's example of drying clothes on a line (382b18-22). Here it is claimed that wet clothes are so dried by cooling, and so by the action of the cold, but not directly by the action of the cold. The cold serves to 'drive out' the heat internal to the wet clothes, which carries along the moisture. The cold also can, paradoxically, heat or burn things, by surrounding some of the hot, and so concentrating it (see *Meteor*. 2.9).

142. Lit., 'gathered into one'.

143. See *Meteor*. 2.9, and Alexander's commentary on it, for the role of the cold in the production of thunder and lightning.

144. See *Probl*. 24.2, 13, where it is asked why in general the inner parts of things are warmer in the winter. See also Theophrastus, Fortenbaugh et al., fr. 173 (= Plutarch, *Aetia Physica* 13, 915B); *de Igne* 16; Antigonus, *Hist. Mir.* 144; Pliny, *Nat. Hist.* 31,50.

145. Excising 'the' (*ta*) prior to 'moist' in 203,6.

146. A close variant of 382b10-11.

147. 382b14-15.

148. The preceding two sentences are meant to distinguish forms of water, which are characterised by their lack of sediment, from sticky things, which also seem to lack sediment, because the sediment is unable to separate from the sticky thing itself.

149. It may sound strange to claim that the hot can evaporate. Perhaps evaporation is more properly ascribed to the whole process, the heat and the moisture exiting together, and not ascribed to the exit of heat alone. But see 206,30-1 below where the hot alone is said to evaporate.

150. See 191,11f. for an account of why boiled things dry out.

151. I propose emending this to say 'by dry-hot, or cold'. See next note for a detailed explanation of the reasons.

152. This is a puzzling passage, commenting on an equally puzzling passage (382b33f.). This text in Aristotle is rendered quite differently in the assorted manuscripts. Fobes and Lee read the text so that everything is solidified by dry-hot or cold (*thermôi xerôi ê psukhrôi*). Aristotle goes on to claim, in the preferred text, that things are solidified by hot or cold, and that those which solidify due to dry-hot are dissolved by water, which is the moist-cold, while those which solidify owing to cold are dissolved by fire – that is the hot. Aristotle, in all the manuscripts, moves from talking of dry-hot as a cause of solidification, to cold *simpliciter* as the other cause (with fire as the cause of the opposed dissolving seemingly only due to the hot). Yet at 383a1 Aristotle glosses his view on what causes solidification (dry-hot or cold) by hot or cold. It is clear he means dry-hot or cold, for in the very next line he returns to talk of dry-hot, when running through these two options. Alexander is at pains to explain this. He does not want one to think that the hot alone could cause solidification. Perhaps he read this passage in Aristotle like many of the manuscripts which *do* isolate the hot as a lone cause of solidification, and so thinks that this claim needs explaining. Yet there still is a problem here. Aristotle, in our best texts, says dry-*hot* or cold, while Alexander has hot or dry-*cold*. Since it is clear that Alexander thinks Aristotle means dry hot, there are two explanations. First, Alexander does mean 'dry-hot or dry-cold' by '*thermôi ê psukhrôi xerôi*' (this is Coutant's interpretation). Alternatively, Alexander's text of Aristotle might read in a way that isolates the hot alone as a cause, or even the hot alone as a cause, and dry-cold as the other. Yet there is a further problem here. What now of the dry-cold? This is not found in any Aristotle manuscript, and seems philosophically out of place. I propose the following resolution. Alexander wrote 'by dry-hot or cold', and either there is a misprint in *CAG*, or all scribes have simply missed this fact. My hunch is confirmed, I think, by 205,9-10, where, in a passage referring back to this one, Alexander states, 'Having said that solidification is by dry-hot or cold ...'. This reading stands in all manuscripts, and cannot be read so as to allow taking 'dry' with both 'hot' and 'cold'.

153. This passage's extreme repetitiveness may be accounted for by Alexander's desire to explain why Aristotle states that boiled honey is solidified by water. It is clear that what is meant is that *already* boiled honey is hardened again by being immersed in cold water, due to the coldness, not the moisture, of the water.

154. This, I take it, helps explain 204,28. See n. 152 above.

155. Alexander's point (and Aristotle's at 383a7-8) is that some power cannot affect something one way at one time, and affect it (or something appropriately similar) in the opposite way at another time, *all things being equal*, and not that, say, fire cannot at the same time, and in both cases due to its heat, melt the ice next to it, and solidify the clay placed upon it, and so the implied 'them' in 205,15.

156. This is a rather convoluted *reductio*. Assuming that something solidified out of water is dissolved by the heat (alone) of fire, the argument proves that the fire cannot also be the cause of such things' solidification. For if one claims that it is the dryness of the fire that causes the initial solidification, then one would have to claim that the moist would be the cause of their dissolution. But this is impossible, since only the hot dissolves, so fire cannot be the cause both of the dissolution of something solidified out of water, and this thing's solidification.

157. Here 'roasting' is akin to 'baking', since we do say that pottery when fired is baked.

158. The point being that things which thicken in the process of being solidified must be both a mixture of earth and water, and moist.

159. 383a21.

160. A close paraphrase of 383a26-7. The supplied clause is taken from 207,1 below.

161. This passage is obscure. Clay solidified by cold may be frozen clay (or Lee's frozen mud). However, I suspect that 'room temperature' clay (which is solid [yet soft] due to the cold) is at issue here. Aristotle claims that upon being heated clay *melts* before hardening (thereby explaining the frequent distortions of pottery when fired). This suggests that the heating at issue is that which takes place in the kiln firing of pottery. Perhaps the normal state in which clay was found had it being quite moist and dissolved, this being its normal nature. Clay which has the consistency to be worked would be 'unnatural' in so far as it lacks sufficient moisture. One should note that at 206,25 Alexander claims only that *the moisture* in clay dissolves in the firing of clay, not the clay itself.

162. This is an interesting passage for a number of reasons. First, it gives us an important clue as to ancient steel-making techniques. This passage (at least) seems to claim that iron ore was not actually melted in the process of steel making, but only softened. Second, the passage in Aristotle this is commenting on (383a29-32, 'Those that solidify because of cold and the carrying away by evaporation of all the hot [in them] are indissoluble except by extreme heat, but can be softened, like iron and horn') seems to be referred to at *GA* 743a1-15, which supports the authenticity of *Meteor.* 4.

163. The term *keras* (horn) includes things like hoofs, or perhaps the substance of hoofs, which soften when heated, but do not melt.

164. Taking *eirgasmenos sidêros* (as in lemma) and *kateirgasmenos sidêros* as synonymous.

165. This passage is obscure. First, one must realise that one is discussing worked (or wrought) iron, which has previously been melted, and so been moist, during its manufacture. It is the moisture of the molten (or, more accurately softened) iron that is at issue. Alexander claims that it is not the very same moisture which returns when this wrought iron is reheated (during the making of steel from the wrought iron). Yet Alexander is bothered by a real worry: how does heating something moisten it? Aristotle does not explain this. He claims that melting is the opposite of solidifying, and that both processes of solidifying (by heat and by cold) involve the moisture of a thing leaving. So, melting should involve moisture returning. Alexander claims that (the original) moisture does not in fact return (how could it by means of heating something extremely?), yet this leaves the melting process mysterious. Alexander falls back on Aristotelian modal metaphysics, which is, here at least, not very helpful. If the wrought iron becomes moist, it was potentially moist, and, upon melting, becomes actually moist. Yet *this* is what needs explaining. Alexander *is* able to explain why it is not, in a sense, actually moist all along, for it is not the kind of solid which is solidified moisture (like ice), yet his claims about being created via the exit of hot vapours (those things which are potentially water, for they can condense) is not illuminating.

166. Literally, the 'fire-fighting' stone. It is in fact limestone, which was used as a flux in ancient furnaces during the production of wrought iron. When in close contact with fire limestone releases carbon dioxide and disintegrates into a form where it acts as a flux. It is this disintegration which Aristotle may be describing here.

167. The confusion concerns both the gender and spelling of these names, and whether they pick out the same stones, or different ones. Alexander thinks that two different kinds of stones must be referred to, since Aristotle says that *mulas* melt, yet below he will state that *muliai* do not dissolve. Only if melting is a kind of dissolving is *this* a real point of conflict. In the text of Aristotle (383b12f., 383b7) the difference seems to be that *mulas* are solidified by the cold, while *muliai* are solidified by dry-hot. These stones (or perhaps only one of them) were also used as a flux in the production of wrought iron, and may be forms of lava.

168. A natural compound containing primarily sodium carbonate, often along with sodium bicarbonate, sodium chloride and sodium sulphate, which the Greeks blended with oil to form a soap, and burnt in incense. It was also used as a preservative, and was an important ingredient used in mummification by the Egyptians. See Pliny, *Nat. Hist.* 31, 106ff. for a discussion of nitron other similar compounds.

169. This is a rather compressed argument from a general to a specific case. Since the cold in general chills pneuma, or air, and turns it into water, so too in the case of the action of the cold on the air in olive oil.

170. The composition and properties of olive oil confound Aristotle. He discusses olive oil in *Meteor.* 4.7, and in *GA* 2.2. The reasoning employed in this passage by Alexander, following Aristotle, is an inference based on an observation. Since olive oil and water mixed are thicker than either taken singly (the observation), the addition of water to olive oil thickens it. In addition, olive oil must contain a lot of air since it floats on water. Now we know that the cold can condense air into water, which is a thickening, since water is thicker than air. So olive oil must thicken by the cold due to the air in it condensing into water. Alexander is quick to note rightly that not too much of the air in the olive oil must undergo such condensation, since to be olive oil is to be full of air, and thickening is not a form of corruption, but only an alteration. Olive oil and water, upon being thoroughly emulsified, do form a seemingly 'thick' mixture due to the tensile strength and stable structure of the many air bubbles in the mixture. See *GA* 2.2 (esp. 735bf.) for an account of thickening due to the formation of many small pneumatic bubbles.

171. Here Alexander disagrees with Aristotle (383b28-9), who claims that both fire and the 'passage of time' thicken *and* lighten olive oil. Alexander is here perhaps drawing upon his own observations, or those of someone who disagrees with Aristotle.

172. 383b31-2.

173. The second set of explanations corresponds to the first. The water evaporating is due to the fire, while the change of air into water is due to time.

174. Wines with much sediment will, upon being heated and boiled away, leave behind a mass of solid sediment, this being the solidification of wine by fire.

175. The unfermented, or partially fermented must was, by the process of fermentation, purified, and separated from (lit. cut away from) the assorted impurities and extraneous matter found in the unfermented must.

176. A very close paraphrase of 384a5-6.

177. Aristotle does not explicitly claim that the vapour is water. He would, no doubt, view the change from vapour to water as a substantial coming to be. See also *Meteor.* 2.3, 358b16f. on this same topic.

178. Perhaps Alexander is here being ironic, for he seems uncertain as to what phenomena Aristotle is describing, since Aristotle does seem to have difficulty accounting for the assorted properties of wine and must. Many of the seemingly contradictory properties of wine and must are due to the varying proofs different samples would have.

179. Aristotle made this claim at 382b1. Here Alexander is commenting upon 384a11.

180. See 208,26f. above for an account of the two stages of oil's thickening by chilled air. I thank Prof. Russell for making sense of this passage.

181. These quotes are from 384a10 and 11, respectively.

182. A close paraphrase of 384a16-17.

183. This may refer to the production of sea salt from the evaporation of sea water, or the salt extracted from the 'salt rivers' of the Caspians, Mardians and Armenians, or from assorted salt lakes. The Bactrian rivers, Ochus and Oxus also contain heavy concentrations of salt. Pliny (*Nat. Hist.* 31, 106-15) discusses the many waters which contained nitron.

184. These formations are like stones, but looser in texture, perhaps calcium deposits.

185. The juice of the fig, being particularly acidic, acts as a rennet.

186. See 384a20-2.

187. A close paraphrase of 384a28-9.

188. Reading *tôi 'ikhôri* in 211,21.

189. The sense in which diseased things are like serum and destroyed blood, is that all of these are watery, lacking solid parts (in the case of blood, fibres). The suppressed premises here are that diseased things are watery, and that diseased blood lacks its solidified parts. It is unclear precisely what the fibrous parts of blood are thought by Aristotle to be, other than merely the earthy component. For a fuller discussion of the fibres in blood see *HA* 3.6 and *PA* 650b15f. Plato discusses the fibrous parts of blood at *Tim.* 82Cf. In general, the fibrous parts of blood are those parts which coagulate, which, upon coagulating form ragged, fibrous lumps.

190. Aristotle (384a33) claims that it is nature itself which fails to master and concoct blood in these cases, not the particular nature of the concocting thing.

191. Since there could not exist any such power.

192. See 207,34f.

193. This section is problematic. Alexander is commenting on 384b14-15, a line with a messy textual history, and ambiguous in any case. Alexander thinks that there are two ways to read this line, and he wants to show that on either reading heat (excessive or otherwise) *dissolves* iron. On the first reading, one is concerned with the production of wrought iron from iron ore. Iron ore is first melted, and then cooled, to produce wrought iron. On this reading both the hot and the cold are needed for the eventual production of solidified wrought iron. Since both are needed for its solidification, it appears that no opposed power is available to dissolve it, and so *excessive* heat is needed. The text of the Aldine Aristotle fits this view. On the second reading Alexander thinks that Aristotle claims that iron is not solidified by the hot, but by the cold alone, and so then the hot, being opposed to the cold, would be able to dissolve it. Yet there is still a problem here. However one reads 384b14-15, it claims only that the hot *melts* iron (either during its production, or pre-existing iron), not that it *dissolves* it. Aristotle seems to be explicitly claiming that iron is only melted, and not dissolved, since both the hot and the cold were involved in its production.

194. See 199,8f. above, where Alexander claims that all mixed bodies have structure.

195. 384b30-1, with very slight changes.

196. 384b33-4.

197. Here Alexander sees continuity between *Meteor.* 4 and *Meteor.* 3, similar to that which Olympiodorus sees, although his final opinion is that *Meteor.* 4 is out of place. See Introduction on the question of the place of *Meteor.* 4 in the corpus.

198. Although these are affections (*pathê*) (and so passive, more often associated with matter than form), they are more form-giving in homeomerous things, since homeomerous things are characterised by their affections, as Alexander (as with Aristotle) will go on to explain. In simple things, like homeomers, form is determined primarily by composition, that is, by the proportions of dry and moist that they are composed out of as their matter. I thank Prof. Russell for helping me make sense of this passage.

199. Alexander is making a claim not explicitly found in Aristotle, namely that homeomerous stuffs are themselves passive (like the dry and the moist, which are passive powers), since they are composed out of water and earth. Aristotle claims (385a8-10) that homeomerous bodies have their natures differentiated primarily by means of the different passivities they are characterised by, yet he does not explicitly make the claim which Alexander here makes.

200. A close paraphrase of 385a22-4.

201. This theory assumes an atomism of a sort, namely that there must, as a matter of physical fact, be a minimum size, or 'bulk' of water.

202. 385b1-2. Being 'of water' here means being primarily of water, since honey and the like does contain water (i.e. the moist), just not enough to be solidified via the drying of this moisture. See 384a14-15, where things 'of earth' are compared with those of 'air and water', implying that the opposition is between those of earth *and* water v. those of air *and* water.

203. Here 'all of the moist' may mean all of the manifest moisture.

204. I translate *elata* as 'malleable', and not 'ductile', so to preserve the contrast with *helktos*. It also makes better sense of the examples given, that of iron and horn, for it is hard to see how horn can be drawn, but it can be worked when heated.

205. Alexander seems to be commenting on a version of 385b6-12 different from that in any known manuscripts. His version may be a variant of MSS HN which read in b10: *alla ê helkta hoion neuron himas*. Alexander's comments make best sense if his text of b9-11 read something like: ... *hôsper ho keramos alla estin helkta hoion neuron kai himas ê elata mê onta hudatos kai* (or *alla* as in HN) *malakta puri* ..., as against the preferred: ... *hôsper ho keramos all'ê helkta mê onta dianta ê elata mê onta hudatos* ..., that is, '... as is clay, but is tractile as sinew and leather, *or* is malleable and not [predominantly] of water is softenable by fire ...', as against the preferred '... as is clay, but if they are either tractile (but) not absorbent, *or* malleable and not [predominantly] of water ...'. Having irregular pores might explain the lack of absorbency (which is found in the text of Aristotle, but goes unmentioned by Alexander), since if the pores are too small, no water could enter them.

206. Alexander thinks that there is a problem with wood, which was said by Aristotle (384b15-16) not to be softened. Here Alexander claims only that *certain* woods can be softened when heated. See also 384b16f., where ebony is said to be an 'exception to the rule' that wood floats.

207. Aristotle has only 'melted' (385b12).

208. There is much confusion concerning the meaning of *tektos*. Lee favours 'softenable by water', while Coutant favours 'softenable by wetting'. In fact it means 'capable of being drenched', that is, able to be in the state of holding water internally. Aristotle is distinguishing various ways something might be liquid-like, or water-like. Some things melt, and so become liquid, while others remain solid, yet are made wet through and through (i.e. drenched). Lee and Coutant have been confused by Aristotle's claim (385b17-18) that nothing is *tektos* which does not become softer when soaked. If *tektos* means 'softenable by wetting' (or by water), Aristotle is here presenting us with a bald tautology. What Aristotle in fact thinks is that some things, upon being made wet, are not made wet through and through, but melt. For something to be made wet through and through, for it to be drenched, it must remain solid. Of course, such things are softened as a result of this drenching. Aristotle is setting up a comparison between solids that melt when made wet, and those which are drenched when made wet.

209. Here, where both Alexander and Aristotle talk of melting, not softening, copper is given as an example, not iron as is found earlier (esp. ch. 6). The Greeks were able to melt and cast copper, but not iron.

210. 385b15-16.

211. 385b19-21, with very minor changes. Something with hard pores has pores which do not dissolve when a liquid enters them, but retain their shape and so hold the liquid.

212. 385b22.

213. This clause must refer to what happens to things drenched but not melted. Bodies which have pores which retain moisture are drenched, unlike those which do not have regular, nonintersecting pores, but 'receive water through the whole of themselves', and are so melted.

214. Alexander relates the notions of the hardness or softness of pores to their topological configuration. Interconnecting pores are soft, while parallel pores are hard.

215. From this point on, Alexander is, for once, briefer than Aristotle in his description of the assorted affections of homogeneous bodies. It is true that Aristotle's description of bending and straightening (385b26-386a9) is itself so baroque that Alexander may have thought that a short gloss was all that was called for.

216. Alexander here reads 'can be softened' (*malakta*), not 'soft' (*malaka*) as with some editors, for what is found at 386a20-1. *Malaka* would yield a difficulty if copper was then said, in comparison, to be hard, implying it cannot be softened, for Alexander uses the term 'hard' *simpliciter* for things that are non-impressible (following Aristotle's comment at 382a10 that hard things do not yield inward at their surface, which impressible things must do). Alexander solves this difficulty by reading 'harder' (the comparative form), not simply 'hard', implying that copper is, in a sense, soft, just not as soft as wax.

217. I render *piestos* as 'compressible', as opposed to Coutant's and Lee's 'squeezable', since anything squeezed is, necessarily, compressed (which yields problems for these editors with the term *piletos*, see n. 220 below). There is in the text of Aristotle a problem none that I know of has remarked upon. At 386a25f., stone is said to be 'compressible' (*piesta*), which means simply 'not *easily* moulded' (i.e. not plastic). Yet at 386b10 stone is lumped together with iron and water as things which cannot be compressed. Perhaps Aristotle meant to distinguish two types of stone. However, it is true that the accounts of the features of solids pertaining to their resistance to topological changes which Aristotle presents are confused due to his not having the dual notions of elasticity and inelasticity.

218. 387a11-12. This is the reading of the majority of manuscripts, following four manuscripts of Aristotle. Fobes and Lee prefer 'being tractile as well as being moist or soft', which makes better philosophical sense, since sticky things are not merely soft and moist, they seem to be a *type* of tractile thing. To take examples from Aristotle, wool is merely tractile, while phlegm is also sticky. Grains can be both soft and moist, according to Aristotle, yet they need not be sticky. I prefer 'sticky' to the more standard 'viscous', since below (217,9) such things are said to contract upon being stretched, indicating a notion closer to elasticity than viscosity.

219. I prefer 'stiff' to the more standard 'friable', since not all things which are not sticky or tractile are easily crumbled, wood for instance. Also better opposition with 'shattered' (*euthrausta*) is obtained if *psathura* means stiff, not friable. Actually, this is an odd term. In Aristotle it is applied to water (*Sens*. 441a25) and air (*DA* 419b35), neither of which is stiff or friable. The term seems to designate things incapable of changing their shape or holding onto a new shape all on their own. Wood, for instance, cannot be pulled or moulded (it must be cut), while water, for instance, holds its shape only due to its container. In other words, things that are *psathura* are either loose in texture (their parts do not form chains), for example water and air, or have an immutable internal texture, like wood. In any case, 'and easily shattered' is Alexander's addition, and does not seem warranted, since, again, wood is stiff and not easily shattered, unlike, say, ice.

220. *piletos* is often translated as compressible, which I have used to translate *piestos* (which is often rendered as 'squeezable', see n. 217 above). My reasons are as follows: Aristotle clearly claims that things with this affection retain their shape upon being compressed (387a15-16). Yet we normally think of sponges and the like as things which can be compressed, since the act of squeezing something is a compression of it, and we do not think that things which are compressible must retain the shape they have while compressed. Aristotle seems to subdivide the impressible into three sub-sets: (1) the compressible and mouldable (the plastic), (2) the non-compressible (iron?), (3) the compressible and not mouldable (stone, wood, wool, sponge, although there are problems with the status of wood). The squashable seems to cut across these divisions, for some of the compressible and not mouldable are squashable while sponges and the like return to their original shape upon no longer being compressed by something. Equally, all plastic things are squashable, since they are both compressible and retain their compressed shape once the compressing has ceased. Again, what Aristotle is after is something like the dual notions of elasticity and inelasticity, the squashable being things that can be compressed, but are inelastic.

221. Alexander's gloss here on 387a17-21 seems to state that it is the moisture that combusts in combustibles, while Aristotle merely claims that the moisture cannot master the incoming fire.

222. The sense of *eis* 'into' here is that of becoming air or wind.

223. This is close to 387a24-5. Alexander substitutes *hugrantikē* for Aristotle's unique *diantikē*. Both terms mean the ability to wet or moisten something, and do not mean, as they are normally taken, to *be* wet or moist, or even the capacity to be wet or moist. What distinguishes vapours from fumes is that vapours moisten things (as with vaporisers), while fumes discolour things, by leaving dry (earthy) residues on that which they come in contact with (as with clothes made sooty by fumes).

224. This passage, and those that follow, are confused, both here and in Aristotle. I take the sense to be the following. Things that give off vapours eject their moisture as either air or wind which themselves are capable of moistening. But the things emitting the vapours are not consumed. An example might be the grilling of meat, where the meat gives off moist vapours, while not itself being consumed. Things that give off fumes, on the other hand, are, along with their moisture, consumed in the process. Both fuming and vaporising is accomplished by the action of heat via fire, yet fumes and vapours differ. Fumable things, like things which give off vapours, do lose their moisture, emitting it 'into' (i.e. as) air, yet not by being burned, as with things which give off vapours, but merely through the passage of time. However, this emission of moisture by time of fumable things still differs from vaporising (called, confusingly, that produced by fire), in that the emissions are not moist, nor do they become wind, only air. I also read Aristotle as saying this (*contra* Lee), and would read 387a26-7 as saying, 'But the fumable can emit [its moisture] into air by the passage of time, and then dry up or become earth.' Examples might be the petrifying of wood, the drying up of foods that fume, etc.

225. Deleting '*hê ... tou hugrou*' in 217,26.

226. 387a29-30. Coutant states, 'Alexander follows the lines *estin ... mêkos*, 387a29f., which are suspect to me, for the definition of wind as air in motion is contradicted by *Meteor.* 1.13, 349a16ff. and 2.4, 360a27ff., where air and wind are rigidly distinguished. Perhaps a careless reading of *Meteor.* 1.13, where the view of others is given led to an interpolation here.'

227. Alexander seems to realise the problem with Aristotle here claiming that wind is a continuous longitudinal flow of air. Elsewhere in the *Meteor.* Aristotle denies that wind is such (349a16, 360a27), and in the latter of these two passages, he claims that wind must have a definite source, and not be any old movement of air. He seems to wish to distinguish legitimate meteorological phenomena from other movements of air (say that resulting from the slamming of a door). Alexander adds this requirement to the text as here found ('receiving its flow from a particular source'), which may in fact be the best way to take the Aristotelian text.

228. Aristotle makes it clear (387b1-3) that calling these all woody is simply due to the lack of an accurate common name.

229. 387b6. Alternatively, the final clause may be, 'of fatty things, fatty steam'.

230. Presumably the reason why oils do not boil or thicken when fuming is that both boiling and thickening were defined as processes where the moist inherent in a thing is drawn out (in boiling by the heat in the water, in thickening by either heat or cold). Fuming involves the moist *together* with the dry being ejected, which yields the consumption of the fuming body, not its thickening or boiling. There are problems with this account, since both the moist and dry are thought to be material elements, and so the ejection of either should (perhaps) yield the diminution of the fuming, boiling, or thickening things. In addition, the fact that olive oil does not boil was said to be due to its viscosity, which prevents the moisture from evaporating. Perhaps fumable things in fuming eject long chain-like bits of their matter (dry and moist) which could be broken by extreme heat, yielding their diminution.

231. 387b8-9.

232. See Theophrastus (*de Causis Plant.* 6.4.1) where Theophrastus debates whether 'winy flavour' should be added to the list of seven basic flavours.

233. The properties of must are a bit mysterious. If it had a greater proof than ordinary wine (perhaps the Falernian new wines with a high sugar content which Pliny discusses), one could explain why they burn like olive oil. Pliny claims that such wines would burn due to their high alcohol content. However must is also claimed not to intoxicate, which suggests that it is the low alcohol pre-fermentation grape juice.

234. Aristotle (387b22-3) states only that these things are less inflammable, but are still somewhat inflammable.

235. 387b24. This is an odd claim, for it is unclear what 'containing' smoke means. The construction suggests composition out of earth and smoke, which makes no sense. The claim that flame is burning smoke may give some credence to the model which has it that bodies 'contain' smoke, yet this passage too is odd, and in any case this claim could be accounted for by a model which has it that inflammable things produce smoke which, upon burning, itself produces flame. One would prefer something like 'Those inflammables which are mainly of earth, or smoke (verbal form from *kapnoomai* or *kapuô*), produce charcoal', but there is no manuscript precedent for this in Aristotle or Alexander. Perhaps we have here a modal 'in', things which have smoke 'in' them are things which can produce smoke. Alternatively, as suggested by Prof. Russell, *kapnou* may be corrupt for *hugou* (moist).

236. 388a2.

237. This talk of the production of fire is as if the dry, being matter, upon being informed by the hot, becomes fire, an element.

238. 388a8-9. The 'of moist things' clause is needlessly repetitive, and a bit forced as translated. It is excised by both Lee and Webster.

239. 388a12-13.

240. This may be a reference to 385a3, where auditory properties of things were mentioned, along with others, in a list of how homeomerous bodies differ from one another.

241. Here I follow Coutant and MS B in reading *homoiomeresin*, not *anomoiomeresin*, since it makes more sense given what Aristotle says at 388a16f.

242. Alexander is here commenting on 388a21-6. He is reading, it seems, along with MS H, *hôs to* in line 22 against the preferred, yet odd, *hôste*.

243. Here, perhaps surprisingly, moist and dry are characterised as powers. Yet one should recall that *Meteor.* 4.8 discusses passive affections, and what they do – how, in a sense, they are active.

244. This distinction is confusing. It seems that a simple body is one composed of water (and so moist), while mixed bodies are composed of water and earth, and are so thickened by cold. Yet, of course, all things are actually composed of all four elements. Here composition is being thought of only in terms of the affections of the elements which some given homeomerous body most manifests.

245. 388a27-8.

246. 388a30-1.

247. The point being that milk will not evaporate totally, but will partially evaporate and partially curdle, leaving behind a solid (earthy) residue.

248. 388a31.

249. Ideler, following Vicomercato, conjects 'honey' instead of 'wood' for what is found at 388a31 (although below [220,12] Alexander will characterise honey as being of earth and *water*). Aristotle himself says that wood is of earth and air at 384b15-16. See 212,12, for Alexander's comment on this. Yet *Meteor*. 4.8 claims that homeomers, like wood, are composed out of water and earth. See 213,10 for Alexander on this. Perhaps one can accept the text as it stands by recognising the distinction between what things are composed out of as their matter, which is the dry and moist, and so earth and water, and what things are composed out of in the sense either of the actual things combined in producing them (often their proximate matter), or the powers they most manifest (which include, therefore, the active elements).

250. 388a31-2.

251. This paragraph is Alexander's commentary on a most difficult passage in Aristotle, 388b11-18. Alexander seems to ignore the 'composed of both' and the 'have [in them as constituents] both' clauses in 388b18. This 'dissolves' the ambiguity in the text of Aristotle as to what all the occurrences of 'both' refer to. For Alexander it is the 'hot and the cold' throughout, as he makes clear. Yet here he avoids the issue as to whether bodies are *composed* out of these.

252. 388b18.

253. No mention of this particular river is found in the text of *Meteor*. It is mentioned in *On Marvellous Things Heard*, 836a30. Coutant suggests that this river may be the Elbe or the Vistula, both amber-rich rivers. It is also sometimes identified with the Po; see Herodotus 3.115.

254. The point seems to be that honey is solidified when dropped in cold water due to the ejection of both the hot and the moist in it. Therefore '*hupomenei*' must be synonymous with 'solidify'. Perhaps 'stands firm' would capture the intended meaning, although the term is certainly idiomatic.

255. The term *êlektron* is ambiguous, since it names both that which we call amber, and a precious alloy of gold and silver.

256. See n. 253 above.

257. Alexander is here careful to explain the sense in which these things are solidified by both the hot and the cold. These things are solidified primarily (*mallon*) by the action of the cold. Yet the hot also contributes, since the cold drives off the hot, and the hot carries along with it the moisture, and it is this dual exit of the hot with the moist which yields solidification. Since the cold on its own cannot drive off the moist, the hot also is a cause of the solidification, and so such things are, properly speaking, solidified by both the hot and the cold.

258. Alexander is here clear that the issue is whether something is *composed* of earth, water, or both, reading 'both' for the 'or have more than one in common' found in the text (389a5-6). He also claims that the causes of thickening are the hot, the cold, or both, against 'fire, or cold, or both' found in the text (389a6-7). Alexander's reading, which, although without manuscript precedent is in ways preferable, supports the model which has it that only the passive powers are material, and the active powers are formal. The substitution of earth and water for the passive powers of dry and cold is reasonable, since we are more used to thinking of matter as being a stuff, as opposed to a power or affection, while we have no difficulty in talking of powers as being active, and so playing causal roles. Alexander does not comment on the end of 4.10 (389a7-23), having incorporated much of what is found there into earlier parts of his commentary.

259. 389a24-5, the opening line of 4.11.

260. For this definition of putrefaction see 379a16-18, and 183,34-184,3 above.

For the earlier account of spontaneous generation, see 379b6-9, and 185,8-11 above.

261. Note that here heat is the cause of the concoction, it is not itself being concocted, the difference indicated by an unusual use of *peptô* in the active, not passive, voice.

262. 389b9-11.

263. 389b15.

264. This is the first and only mention of anything other than earth or water (here air) being matter. It parallels its only occurrence in the text of Aristotle (389b17).

265. The point here is not that cold-solidified things out of water are *burnt* when heated, but that they in turn can *burn* most severely upon having been heated (and sometimes therefore liquefied, as in ice becoming boiling water). According to Aristotle and Alexander, sticking one's hand in a flame is not nearly as bad as plunging it into hot water, which, in turn, burns less than the worst fate, touching heated rocks (perhaps coals are in mind here).

266. *logos* here, and in the following, means 'definition'.

267. 389b27-8.

268. 389b27.

269. That is, the anomeomers have as their matter the homeomers that they are composed out of, but they do not have the same form as these homeomers.

270. Reading the '*kai*' here as epexegetic, since Alexander takes *ousia* and *to einai* as synonymous. *Ousia* is therefore best rendered by 'being', although I shall continue to use 'substance' in order to mark its distinction from *to einai*.

271. *ei ge* here having the sense of 'since'.

272. Here *horizetai* clearly means 'defines', and not 'determines', as it has been previously translated throughout this work. There is a problem as to how to translate it in the rest of this chapter. However, the two meanings, 'to determine' and 'to define', are, in fact, very closely related. For more on this see Introduction.

273. These are sculptured representations of pipes, and cannot function as instruments.

274. That is, stone flutes are called flutes but equivocally, similarly with other things that cannot fulfil their proper function.

275. In other words, the elements are the matter of compounds. Note that Alexander does not consider the elements themselves to be 'pure' matter, only to be closest to matter.

276. This passage is obscure, since eyes are not themselves homeomerous. Perhaps we should emend to [*an*]*omoiomerôn*, that is, '... compounded out of certain *an*omeomerous bodies ...'.

277. Meaning, I take it, those which have underlying matter itself without form.

278. See Introduction, n. 51.

279. 390a9.

280. This is an important claim, for it is recognition, both by Aristotle and Alexander, that there is a function (and so, I take it a formal aspect to) the elements. See 390a15, where this is explicitly said to be true for fire. 224,30f. above also makes this claim explicit, namely that the elements, just like anomeomers, have a function and a 'for the sake of which'.

281. From the grammar of this sentence it is unclear whether the parts of animals are here said to be defined by their powers in general (both passive and active), or just by their passive powers, as with earth. *Meteor*. 4.8 distinguishes homeomers via their differing passive powers. Yet it does not there claim that they are *defined* via these, only that they can be so differentiated. It would be odd to define the parts of animals this way, since a hunk of a corpse's flesh may have the same passive powers as when the person was alive, differing only with regards to its active powers. Yet we know that the 'dead' flesh is called flesh equivocally only. Alexander, in his commentary on 4.8, claims that the passive powers are more form giving. This seems true for homeomerous stuffs in isolation, but not those things, like animal parts, which form part of a system, and have functions above and beyond those which follow directly from being a certain kind of matter.

282. Here Alexander reads 'from (*apo*) these', as opposed to 'by (*hupo*) these' as found in the preferred manuscripts.

283. This term is not found earlier in Alexander, or in Aristotle. It may be a general term meant to include a number of the passive powers discussed in 4.8-9.

284. This is also the first occurrence of this term, which seems to incorporate both things that break and those that shatter.

285. 390b7-8.

286. This is not to deny that anomeomers, say a face and a hand, may differ from each other with respect to assorted passive powers, but only that this is not how they are *said* to differ, since these features do not enter into the definition or being of the anomeomers. It is not in virtue of these that anomeomers are said to differ, they may in fact differ with respect to these due to the real explanation of their difference, their different functions. It is because a hand and a face are for certain ends that they may have to, say, have different degrees of hardness.

287. I render both occurrences of *kai* as being epexegetic. It is difficult to see otherwise what the substrate could be other than the matter, and Alexander does seem to equate *logos* in its technical sense of definition, with form. In fact, in Aristotle (390b18) we find both *hulê* (matter) and *logos*. Alexander is expanding upon these two.

288. 403a29f.

289. 390b17-18.

290. This is a deeply puzzling passage. Aristotle (390b17-18) claims that we know the 'why it is' and the 'what it is' when we grasp the material cause (*hulên*), or the definition (*logon* – the formal cause) of a thing, specifically of its generation and destruction. There is no mention at all of final causes. In addition, Aristotle claims that we know both the 'why it is' and the 'what it is' when we grasp either the material or formal cause. The manuscript traditions of both the Aristotle and the Alexander are of no help. I have only a conjecture. Alexander drops talk of final causation immediately. When he sums up (the 'hence' [*hôste*] clause) he returns to talk of material and formal causation. We also know that he equates the *logos* of something with its formal cause (224,21-2), and so if his text of Aristotle read '*logon*', one would expect him to equate this with the formal, not the final cause. Therefore, I propose emending the '*tên te telikên*' in 227,8 to '*tên te eidikên*'. Yet why then does Alexander separate the two causes (formal and material), associating the formal with 'why it is', and the material with 'what it is', while Aristotle claims you can know *both* of these from either cause (itself an odd claim)? We can make sense of this if the text of Aristotle which Alexander had read '*ean ton logon kai tên hulên ekhômen*' in 390b17-18, or perhaps even '*ean ton logon ê tên hulên ekhômen*' (dropping the first '*ê*' as in MS E and the version of Henricus Aristippus). That is, 'if we grasp the formal and the material cause', or 'if we grasp the formal or material cause', as against the preferred 'if we grasp either the material or the formal cause', for this would warrant Alexander's correspondence between the formal cause and the 'why it is', and the material cause and the 'what it is'. Alternatively, this is all just a bit of rather forced and unhelpful exegesis!

    291. 390b19.

Appendix
The Commentators*

The 15,000 pages of the Ancient Greek Commentaries on Aristotle are the largest corpus of Ancient Greek philosophy that has not been translated into English or other European languages. The standard edition (*Commentaria in Aristotelem Graeca*, or *CAG*) was produced by Hermann Diels as general editor under the auspices of the Prussian Academy in Berlin. Arrangements have now been made to translate at least a large proportion of this corpus, along with some other Greek and Latin commentaries not included in the Berlin edition, and some closely related non-commentary works by the commentators.

The works are not just commentaries on Aristotle, although they are invaluable in that capacity too. One of the ways of doing philosophy between A.D. 200 and 600, when the most important items were produced, was by writing commentaries. The works therefore represent the thought of the Peripatetic and Neoplatonist schools, as well as expounding Aristotle. Furthermore, they embed fragments from all periods of Ancient Greek philosophical thought: this is how many of the Presocratic fragments were assembled, for example. Thus they provide a panorama of every period of Ancient Greek philosophy.

The philosophy of the period from A.D. 200 to 600 has not yet been intensively explored by philosophers in English-speaking countries, yet it is full of interest for physics, metaphysics, logic, psychology, ethics and religion. The contrast with the study of the Presocratics is striking. Initially the incomplete Presocratic fragments might well have seemed less promising, but their interest is now widely known, thanks to the philological and philosophical effort that has been concentrated upon them. The incomparably vaster corpus which preserved so many of those fragments offers at least as much interest, but is still relatively little known.

The commentaries represent a missing link in the history of philosophy: the Latin-speaking Middle Ages obtained their knowledge of Aristotle at least partly through the medium of the commentaries. Without an appreciation of this, mediaeval interpretations of Aristotle will not be understood. Again, the ancient commentaries are the unsuspected source of ideas which have been thought, wrongly, to originate in the later mediaeval period. It has been supposed, for example, that Bonaventure in the thirteenth century invented the ingenious arguments based on the concept of infinity which attempt to prove the Christian view that the universe had a beginning. In fact, Bonaventure is merely repeating arguments devised

* Reprinted from the Editor's General Introduction to the series in Christian Wildberg, *Philoponus Against Aristotle on the Eternity of the World*, London and Ithaca, N.Y., 1987.

by the commentator Philoponus 700 years earlier and preserved in the meantime by the Arabs. Bonaventure even uses Philoponus' original examples. Again, the introduction of impetus theory into dynamics, which has been called a scientific revolution, has been held to be an independent invention of the Latin West, even if it was earlier discovered by the Arabs or their predecessors. But recent work has traced a plausible route by which it could have passed from Philoponus, via the Arabs, to the West.

The new availability of the commentaries in the sixteenth century, thanks to printing and to fresh Latin translations, helped to fuel the Renaissance break from Aristotelian science. For the commentators record not only Aristotle's theories, but also rival ones, while Philoponus as a Christian devises rival theories of his own and accordingly is mentioned in Galileo's early works more frequently than Plato.[1]

It is not only for their philosophy that the works are of interest. Historians will find information about the history of schools, their methods of teaching and writing and the practices of an oral tradition.[2] Linguists will find the indexes and translations an aid for studying the development of word meanings, almost wholly uncharted in Liddell and Scott's *Lexicon*, and for checking shifts in grammatical usage.

Given the wide range of interests to which the volumes will appeal, the aim is to produce readable translations, and to avoid so far as possible presupposing any knowledge of Greek. Notes will explain points of meaning, give cross-references to other works, and suggest alternative interpretations of the text where the translator does not have a clear preference. The introduction to each volume will include an explanation why the work was chosen for translation: none will be chosen simply because it is there. Two of the Greek texts are currently being re-edited – those of Simplicius *in Physica* and *in de Caelo* – and new readings will be exploited by

1. See Fritz Zimmermann, 'Philoponus' impetus theory in the Arabic tradition'; Charles Schmitt, 'Philoponus' commentary on Aristotle's *Physics* in the sixteenth century', and Richard Sorabji, 'John Philoponus', in Richard Sorabji (ed.), *Philoponus and the Rejection of Aristotelian Science* (London and Ithaca, N.Y. 1987).

2. See e.g. Karl Praechter, 'Die griechischen Aristoteleskommentare', *Byzantinische Zeitschrift* 18 (1909), 516-38 (translated into English in R. Sorabji (ed.), *Aristotle Transformed: the ancient commentators and their influence* (London and Ithaca, N.Y. 1990); M. Plezia, *de Commentariis Isagogicis* (Cracow 1947); M. Richard, '*Apo Phônês*', *Byzantion* 20 (1950), 191-222; É. Evrard, *L'Ecole d'Olympiodore et la composition du commentaire à la physique de Jean Philopon*, Diss. (Liège 1957); L.G. Westerink, *Anonymous Prolegomena to Platonic Philosophy* (Amsterdam 1962) (new revised edition, translated into French, Collection Budé; part of the revised introduction, in English, is included in *Aristotle Transformed*); A.-J. Festugière, 'Modes de composition des commentaires de Proclus', *Museum Helveticum* 20 (1963), 77-100, repr. in his *Études* (1971), 551-74; P. Hadot, 'Les divisions des parties de la philosophie dans l'antiquité', *Museum Helveticum* 36 (1979), 201-23; I. Hadot, 'La division néoplatonicienne des écrits d'Aristote', in J. Wiesner (ed.), *Aristoteles Werk und Wirkung* (Paul Moraux gewidmet), vol. 2 (Berlin 1986); I. Hadot, 'Les introductions aux commentaires exégétiques chez les auteurs néoplatoniciens et les auteurs chrétiens', in M. Tardieu (ed.), *Les règles de l'interprétation* (Paris 1987), 99-119. These topics are treated, and a bibliography supplied, in *Aristotle Transformed*.

translators as they become available. Each volume will also contain a list of proposed emendations to the standard text. Indexes will be of more uniform extent as between volumes than is the case with the Berlin edition, and there will be three of them: an English-Greek glossary, a Greek-English index, and a subject index.

The commentaries fall into three main groups. The first group is by authors in the Aristotelian tradition up to the fourth century A.D. This includes the earliest extant commentary, that by Aspasius in the first half of the second century A.D. on the *Nicomachean Ethics*. The anonymous commentary on Books 2, 3, 4 and 5 of the *Nicomachean Ethics*, in *CAG* vol. 20, is derived from Adrastus, a generation later.[3] The commentaries by Alexander of Aphrodisias (appointed to his chair between A.D. 198 and 209) represent the fullest flowering of the Aristotelian tradition. To his successors Alexander was The Commentator *par excellence*. To give but one example (not from a commentary) of his skill at defending and elaborating Aristotle's views, one might refer to his defence of Aristotle's claim that space is finite against the objection that an edge of space is conceptually problematic.[4] Themistius (*fl*. late 340s to 384 or 385) preferred paraphrasing Aristotelian treatises to competing with the larger commentaries of his predecessors, notably Alexander of Aphrodisias.[5] In fact, the Neoplatonists were to introduce new dimensions into commentary. Themistius' own relation to the Neoplatonist as opposed to the Aristotelian tradition is a matter of controversy,[6] but it would be agreed that his commentaries show far less bias than the full-blown Neoplatonist ones. They are also far more informative than the designation 'paraphrase' might suggest, and it has been estimated that Philoponus' *Physics* commentary draws silently on Themistius six hundred times.[7] The pseudo-Alexandrian commentary on *Metaphysics*

3. Anthony Kenny, *The Aristotelian Ethics* (Oxford 1978), 37, n.3: Paul Moraux, *Der Aristotelismus bei den Griechen*, vol. 2 (Berlin 1984), 323-30.

4. Alexander, *Quaestiones* 3.12, discussed in my *Matter, Space and Motion* (London and Ithaca, N.Y. 1988). For Alexander see R.W. Sharples, 'Alexander of Aphrodisias: scholasticism and innovation', in W. Haase (ed.), *Aufstieg und Niedergang der römischen Welt*, part 2 *Principat*, vol. 36.2, *Philosophie und Wissenschaften* (1987).

5. Themistius *in An. Post.* 1,2-12. See H.J. Blumenthal, 'Photius on Themistius (Cod. 74): did Themistius write commentaries on Aristotle?', *Hermes* 107 (1979), 168-82.

6. For different views, see H.J. Blumenthal, 'Themistius, the last Peripatetic commentator on Aristotle?', in Glen W. Bowersock, Walter Burkert, Michael C.J. Putnam, *Arktouros*, Hellenic Studies Presented to Bernard M.W. Knox (Berlin and N.Y., 1979), 391-400; E.P. Mahoney, 'Themistius and the agent intellect in James of Viterbo and other thirteenth-century philosophers: (Saint Thomas Aquinas, Siger of Brabant and Henry Bate)', *Augustiniana* 23 (1973), 422-67, at 428-31; id., 'Neoplatonism, the Greek commentators and Renaissance Aristotelianism', in D.J. O'Meara (ed.), *Neoplatonism and Christian Thought* (Albany N.Y. 1982), 169-77 and 264-82, esp. n. 1, 264-6; Robert Todd, introduction to translation of Themistius *in DA* 3.4-8, in *Two Greek Aristotelian Commentators on the Intellect*, trans. Frederick M. Schroeder and Robert B. Todd (Toronto 1990).

7. H. Vitelli, *CAG* 17, p. 992, s.v. Themistius.

6-14, of unknown authorship, has been placed by some in the same group of commentaries as being earlier than the fifth century.[8]

By far the largest group of extant commentaries is that of the Neoplatonists up to the sixth century A.D. Nearly all the major Neoplatonists, apart from Plotinus (the founder of Neoplatonism), wrote commentaries on Aristotle, although those of Iamblichus (*c.* 250–*c.* 325) survive only in fragments, and those of three Athenians, Plutarchus (died 432), his pupil Proclus (410–485) and the Athenian Damascius (*c.* 462–after 538), are lost.[9] As a result of these losses, most of the extant Neoplatonist commentaries come from the late fifth and the sixth centuries and a good proportion from Alexandria. There are commentaries by Plotinus' disciple and editor Porphyry (232–309), by Iamblichus' pupil Dexippus (*c.* 330), by Proclus' teacher Syrianus (died *c.* 437), by Proclus' pupil Ammonius (435/445–517/526), by Ammonius' three pupils Philoponus (*c.* 490 to 570s), Simplicius (wrote after 532, probably after 538) and Asclepius (sixth century), by Ammonius' next but one successor Olympiodorus (495/505–after 565), by Elias (*fl.* 541?), by David (second half of the sixth century, or beginning of the seventh) and by Stephanus (took the chair in Constantinople *c.* 610). Further, a commentary on the *Nicomachean Ethics* has been ascribed to Heliodorus of Prusa, an unknown pre-fourteenth-century figure, and there is a commentary by Simplicius' colleague Priscian of Lydia on Aristotle's successor Theophrastus. Of these commentators some of the last were Christians (Philoponus, Elias, David and Stephanus), but they were Christians writing in the Neoplatonist tradition, as was also Boethius who produced a number of commentaries in Latin before his death in 525 or 526.

The third group comes from a much later period in Byzantium. The Berlin edition includes only three out of more than a dozen commentators described in Hunger's *Byzantinisches Handbuch*.[10] The two most important are Eustratius (1050/1060–*c*.1120), and Michael of Ephesus. It has been suggested that these two belong to a circle organised by the princess

8. The similarities to Syrianus (died *c.* 437) have suggested to some that it predates Syrianus (most recently Leonardo Tarán, review of Paul Moraux, *Der Aristotelismus*, vol.1 in *Gnomon* 46 (1981), 721-50 at 750), to others that it draws on him (most recently P. Thillet, in the Budé edition of Alexander *de Fato*, p. lvii). Praechter ascribed it to Michael of Ephesus (eleventh or twelfth century), in his review of *CAG* 22.2, in *Göttingische Gelehrte Anzeiger* 168 (1906), 861-907.

9. The Iamblichus fragments are collected in Greek by Bent Dalsgaard Larsen, *Jamblique de Chalcis, Exégète et Philosophe* (Aarhus 1972), vol. 2. Most are taken from Simplicius, and will accordingly be translated in due course. The evidence on Damascius' commentaries is given in L.G. Westerink, *The Greek Commentaries on Plato's Phaedo*, vol. 2, Damascius (Amsterdam 1977), 11-12; on Proclus' in L.G. Westerink, *Anonymous Prolegomena to Platonic Philosophy* (Amsterdam 1962), xii, n. 22; on Plutarchus' in H.M. Blumenthal, 'Neoplatonic elements in the de Anima commentaries', *Phronesis* 21 (1976), 75.

10. Herbert Hunger, *Die hochsprachliche profane Literatur der Byzantiner*, vol. 1 (= *Byzantinisches Handbuch*, part 5, vol. 1) (Munich 1978), 25-41. See also B.N. Tatakis, *La Philosophie Byzantine* (Paris 1949).

Anna Comnena in the twelfth century, and accordingly the completion of Michael's commentaries has been redated from 1040 to 1138.[11] His commentaries include areas where gaps had been left. Not all of these gap-fillers are extant, but we have commentaries on the neglected biological works, on the *Sophistici Elenchi*, and a small fragment of one on the *Politics*. The lost *Rhetoric* commentary had a few antecedents, but the *Rhetoric* too had been comparatively neglected. Another product of this period may have been the composite commentary on the *Nicomachean Ethics* (*CAG* 20) by various hands, including Eustratius and Michael, along with some earlier commentators, and an improvisation for Book 7. Whereas Michael follows Alexander and the conventional Aristotelian tradition, Eustratius' commentary introduces Platonist, Christian and anti-Islamic elements.[12]

The composite commentary was to be translated into Latin in the next century by Robert Grosseteste in England. But Latin translations of various logical commentaries were made from the Greek still earlier by James of Venice (*fl. c.* 1130), a contemporary of Michael of Ephesus, who may have known him in Constantinople. And later in that century other commentaries and works by commentators were being translated from Arabic versions by Gerard of Cremona (died 1187).[13] So the twelfth century resumed the transmission which had been interrupted at Boethius' death in the sixth century.

The Neoplatonist commentaries of the main group were initiated by Porphyry. His master Plotinus had discussed Aristotle, but in a very independent way, devoting three whole treatises (*Enneads* 6.1-3) to attacking Aristotle's classification of the things in the universe into categories. These categories took no account of Plato's world of Ideas, were inferior to Plato's classifications in the *Sophist* and could anyhow be collapsed, some

11. R. Browning, 'An unpublished funeral oration on Anna Comnena', *Proceedings of the Cambridge Philological Society* n.s. 8 (1962), 1-12, esp. 6-7.

12. R. Browning, op. cit. H.D.P. Mercken, *The Greek Commentaries of the Nicomachean Ethics of Aristotle in the Latin Translation of Grosseteste, Corpus Latinum Commentariorum in Aristotelem Graecorum* VI 1 (Leiden 1973), ch. 1, 'The compilation of Greek commentaries on Aristotle's Nicomachean Ethics'. Sten Ebbesen, 'Anonymi Aurelianensis I Commentarium in *Sophisticos Elenchos*', *Cahiers de l'Institut Moyen Age Grecque et Latin* 34 (1979), 'Boethius, Jacobus Veneticus, Michael Ephesius and "Alexander" ', pp. v-xiii; id., *Commentators and Commentaries on Aristotle's Sophistici Elenchi*, 3 parts, *Corpus Latinum Commentariorum in Aristotelem Graecorum*, vol. 7 (Leiden 1981); A. Preus, *Aristotle and Michael of Ephesus on the Movement and Progression of Animals* (Hildesheim 1981), introduction.

13. For Grosseteste, see Mercken as in n. 12. For James of Venice, see Ebbesen as in n. 12, and L. Minio-Paluello, 'Jacobus Veneticus Grecus', *Traditio* 8 (1952), 265-304; id., 'Giacomo Veneto e l'Aristotelismo Latino', in Pertusi (ed.), *Venezia e l'Oriente fra tardo Medioevo e Rinascimento* (Florence 1966), 53-74, both reprinted in his *Opuscula* (1972). For Gerard of Cremona, see M. Steinschneider, *Die europäischen Übersetzungen aus dem arabischen bis Mitte des 17. Jahrhunderts* (repr. Graz 1956); E. Gilson, *History of Christian Philosophy in the Middle Ages* (London 1955), 235-6 and more generally 181-246. For the translators in general, see Bernard G. Dod, 'Aristoteles Latinus', in N. Kretzmann, A. Kenny, J. Pinborg (eds), *The Cambridge History of Latin Medieval Philosophy* (Cambridge 1982).

of them into others. Porphyry replied that Aristotle's categories could apply perfectly well to the world of intelligibles and he took them as in general defensible.[14] He wrote two commentaries on the *Categories*, one lost, and an introduction to it, the *Isagôgê*, as well as commentaries, now lost, on a number of other Aristotelian works. This proved decisive in making Aristotle a necessary subject for Neoplatonist lectures and commentary. Proclus, who was an exceptionally quick student, is said to have taken two years over his Aristotle studies, which were called the Lesser Mysteries, and which preceded the Greater Mysteries of Plato.[15] By the time of Ammonius, the commentaries reflect a teaching curriculum which begins with Porphyry's *Isagôgê* and Aristotle's *Categories*, and is explicitly said to have as its final goal a (mystical) ascent to the supreme Neoplatonist deity, the One.[16] The curriculum would have progressed from Aristotle to Plato, and would have culminated in Plato's *Timaeus* and *Parmenides*. The latter was read as being about the One, and both works were established in this place in the curriculum at least by the time of Iamblichus, if not earlier.[17]

Before Porphyry, it had been undecided how far a Platonist should accept Aristotle's scheme of categories. But now the proposition began to gain force that there was a harmony between Plato and Aristotle on most things.[18] Not for the only time in the history of philosophy, a perfectly crazy proposition proved philosophically fruitful. The views of Plato and of Aristotle had both to be transmuted into a new Neoplatonist philosophy in order to exhibit the supposed harmony. Iamblichus denied that Aristotle contradicted Plato on the theory of Ideas.[19] This was too much for Syrianus and his pupil Proclus. While accepting harmony in many areas,[20] they could see that there was disagreement on this issue and also on the issue of whether God was causally responsible for the existence of the ordered

14. See P. Hadot, 'L'harmonie des philosophies de Plotin et d'Aristote selon Porphyre dans le commentaire de Dexippe sur les Catégories', in *Plotino e il neoplatonismo in Oriente e in Occidente* (Rome 1974), 31-47; A.C. Lloyd, 'Neoplatonic logic and Aristotelian logic', *Phronesis* 1 (1955-6), 58-79 and 146-60.

15. Marinus, *Life of Proclus* ch. 13, 157,41 (Boissonade).

16. The introductions to the *Isagôgê* by Ammonius, Elias and David, and to the *Categories* by Ammonius, Simplicius, Philoponus, Olympiodorus and Elias are discussed by L.G. Westerink, *Anonymous Prolegomena* and I. Hadot, 'Les Introductions', see n. 2 above.

17. Proclus in *Alcibiadem 1* p. 11 (Creuzer); Westerink, *Anonymous Prolegomena*, ch. 26, 12f. For the Neoplatonist curriculum see Westerink, Festugière, P. Hadot and I. Hadot in n. 2.

18. See e.g. P. Hadot (1974), as in n. 14 above; H.J. Blumenthal, 'Neoplatonic elements in the de Anima commentaries', *Phronesis* 21 (1976), 64-87; H.A. Davidson, 'The principle that a finite body can contain only finite power', in S. Stein and R. Loewe (eds), *Studies in Jewish Religious and Intellectual History presented to A. Altmann* (Alabama 1979), 75-92; Carlos Steel, 'Proclus et Aristotle', Proceedings of the Congrès Proclus held in Paris 1985, J. Pépin and H.D. Saffrey (eds), *Proclus, lecteur et interprète des anciens* (Paris 1987), 213-25; Koenraad Verrycken, *God en Wereld in de Wijsbegeerte van Ioannes Philoponus*, Ph.D. Diss. (Louvain 1985).

19. Iamblichus ap. Elian *in Cat.* 123,1-3.

20. Syrianus *in Metaph.* 80,4-7; Proclus *in Tim.* 1.6,21-7,16.

physical cosmos, which Aristotle denied. But even on these issues, Proclus' pupil Ammonius was to claim harmony, and, though the debate was not clear cut,[21] his claim was on the whole to prevail. Aristotle, he maintained, accepted Plato's Ideas,[22] at least in the form of principles (*logoi*) in the divine intellect, and these principles were in turn causally responsible for the beginningless existence of the physical universe. Ammonius wrote a whole book to show that Aristotle's God was thus an efficent cause, and though the book is lost, some of its principal arguments are preserved by Simplicius.[23] This tradition helped to make it possible for Aquinas to claim Aristotle's God as a Creator, albeit not in the sense of giving the universe a beginning, but in the sense of being causally responsible for its beginningless existence.[24] Thus what started as a desire to harmonise Aristotle with Plato finished by making Aristotle safe for Christianity. In Simplicius, who goes further than anyone,[25] it is a formally stated duty of the commentator to display the harmony of Plato and Aristotle in most things.[26] Philoponus, who with his independent mind had thought better of his earlier belief in harmony, is castigated by Simplicius for neglecting this duty.[27]

The idea of harmony was extended beyond Plato and Aristotle to Plato and the Presocratics. Plato's pupils Speusippus and Xenocrates saw Plato as being in the Pythagorean tradition.[28] From the third to first centuries B.C., pseudo-Pythagorean writings present Platonic and Aristotelian doctrines as if they were the ideas of Pythagoras and his pupils,[29] and these forgeries were later taken by the Neoplatonists as genuine. Plotinus saw the Presocratics as precursors of his own views,[30] but Iamblichus went far beyond him by writing ten volumes on Pythagorean philosophy.[31] Thereafter Proclus sought to unify the whole of

21. Asclepius sometimes accepts Syranius' interpretation (*in Metaph.* 433,9-436,6); which is, however, qualified, since Syrianus thinks Aristotle is realy committed willy-nilly to much of Plato's view (*in Metaph.* 117,25-118,11; ap. Asclepium *in Metaph.* 433,16; 450,22); Philoponus repents of his early claim that Plato is not the target of Aristotle's attack, and accepts that Plato is rightly attacked for treating ideas as independent entities outside the divine Intellect (*in DA* 37,18-31; *in Phys.* 225,4-226,11; *contra Procl.* 26,24-32,13; *in An. Post.* 242,14-243,25).

22. Asclepius *in Metaph.* from the voice of (i.e. from the lectures of) Ammonius 69,17-21; 71,28; cf. Zacharias *Ammonius, Patrologia Graeca* vol. 85 col. 952 (Colonna).

23. Simplicius *in Phys.* 1361,11-1363,12. See H.A. Davidson; Carlos Steel; Koenraad Verrycken in n. 18 above.

24. See Richard Sorabji, *Matter, Space and Motion* (London and Ithaca, N.Y. 1988), ch. 15.

25. See e.g. H.J. Blumenthal in n. 18 above.

26. Simplicius *in Cat.* 7,23-32.

27. Simplicius *in Cael.* 84,11-14; 159,2-9. On Philoponus' *volte face* see n. 21 above.

28. See e.g. Walter Burkert, *Weisheit und Wissenschaft* (Nürnberg 1962), translated as *Lore and Science in Ancient Pythagoreanism* (Cambridge Mass. 1972), 83-96.

29. See Holger Thesleff, *An Introduction to the Pythagorean Writings of the Hellenistic Period* (Åbo 1961); Thomas Alexander Szlezák, *Pseudo-Archytas über die Kategorien*, Peripatoi vol. 4 (Berlin and New York 1972).

30. Plotinus e.g. 4.8.1; 5.1.8 (10-27); 5.1.9.

31. See Dominic O'Meara, *Pythagoras Revived: Mathematics and Philosophy in Late Antiquity* (Oxford 1989).

Greek philosophy by presenting it as a continuous clarification of divine revelation[32] and Simplicius argued for the same general unity in order to rebut Christian charges of contradictions in pagan philosophy.[33]

Later Neoplatonist commentaries tend to reflect their origin in a teaching curriculum:[34] from the time of Philoponus, the discussion is often divided up into lectures, which are subdivided into studies of doctrine and of text. A general account of Aristotle's philosophy is prefixed to the *Categories* commentaries and divided, according to a formula of Proclus,[35] into ten questions. It is here that commentators explain the eventual purpose of studying Aristotle (ascent to the One) and state (if they do) the requirement of displaying the harmony of Plato and Aristotle. After the ten-point introduction to Aristotle, the *Categories* is given a six-point introduction, whose antecedents go back earlier than Neoplatonism, and which requires the commentator to find a unitary theme or scope (*skopos*) for the treatise. The arrangements for late commentaries on Plato are similar. Since the Plato commentaries form part of a single curriculum they should be studied alongside those on Aristotle. Here the situation is easier, not only because the extant corpus is very much smaller, but also because it has been comparatively well served by French and English translators.[36]

Given the theological motive of the curriculum and the pressure to harmonise Plato with Aristotle, it can be seen how these commentaries are a major source for Neoplatonist ideas. This in turn means that it is not safe to extract from them the fragments of the Presocratics, or of other authors, without making allowance for the Neoplatonist background against which the fragments were originally selected for discussion. For different reasons, analogous warnings apply to fragments preserved by the pre-Neoplatonist commentator Alexander.[37] It will be another advantage of the present translations that they will make it easier to check the distorting effect of a commentator's background.

Although the Neoplatonist commentators conflate the views of Aristotle with those of Neoplatonism, Philoponus alludes to a certain convention

32. See Christian Guérard, 'Parménide d'Elée selon les Néoplatoniciens', forthcoming.

33. Simplicius *in Phys.* 28,32-29,5; 640,12-18. Such thinkers as Epicurus and the Sceptics, however, were not subject to harmonisation.

34. See the literature in n. 2 above.

35. ap. Elian *in Cat.* 107,24-6.

36. English: Calcidius *in Tim.* (parts by van Winden; den Boeft); Iamblichus fragments (Dillon); Proclus *in Tim.* (Thomas Taylor); Proclus *in Parm.* (Dillon); Proclus *in Parm.*, end of 7th book, from the Latin (Klibansky, Labowsky, Anscombe); Proclus *in Alcib. 1* (O'Neill); Olympiodorus and Damascius *in Phaedonem* (Westerink); Damascius *in Philebum* (Westerink); *Anonymous Prolegomena to Platonic Philosophy* (Westerink). See also extracts in Thomas Taylor, *The Works of Plato*, 5 vols. (1804). French: Proclus *in Tim.* and *in Rempublicam* (Festugière); *in Parm.* (Chaignet); Anon. *in Parm* (P. Hadot); Damascius *in Parm.* (Chaignet).

37. For Alexander's treatment of the Stoics, see Robert B. Todd, *Alexander of Aphrodisias on Stoic Physics* (Leiden 1976), 24-9.

when he quotes Plutarchus expressing disapproval of Alexander for expounding his own philosophical doctrines in a commentary on Aristotle.[38] But this does not stop Philoponus from later inserting into his own commentaries on the *Physics* and *Meteorology* his arguments in favour of the Christian view of Creation. Of course, the commentators also wrote independent works of their own, in which their views are expressed independently of the exegesis of Aristotle. Some of these independent works will be included in the present series of translations.

The distorting Neoplatonist context does not prevent the commentaries from being incomparable guides to Aristotle. The introductions to Aristotle's philosophy insist that commentators must have a minutely detailed knowledge of the entire Aristotelian corpus, and this they certainly have. Commentators are also enjoined neither to accept nor reject what Aristotle says too readily, but to consider it in depth and without partiality. The commentaries draw one's attention to hundreds of phrases, sentences and ideas in Aristotle, which one could easily have passed over, however often one read him. The scholar who makes the right allowance for the distorting context will learn far more about Aristotle than he would be likely to on his own.

The relations of Neoplatonist commentators to the Christians were subtle. Porphyry wrote a treatise explicitly against the Christians in 15 books, but an order to burn it was issued in 448, and later Neoplatonists were more circumspect. Among the last commentators in the main group, we have noted several Christians. Of these the most important were Boethius and Philoponus. It was Boethius' programme to transmit Greek learning to Latin-speakers. By the time of his premature death by execution, he had provided Latin translations of Aristotle's logical works, together with commentaries in Latin but in the Neoplatonist style on Porphyry's *Isagôgê* and on Aristotle's *Categories* and *de Interpretatione*, and interpretations of the *Prior* and *Posterior Analytics*, *Topics* and *Sophistici Elenchi*. The interruption of his work meant that knowledge of Aristotle among Latin-speakers was confined for many centuries to the logical works. Philoponus is important both for his proofs of the Creation and for his progressive replacement of Aristotelian science with rival theories, which were taken up at first by the Arabs and came fully into their own in the West only in the sixteenth century.

Recent work has rejected the idea that in Alexandria the Neoplatonists compromised with Christian monotheism by collapsing the distinction between their two highest deities, the One and the Intellect. Simplicius (who left Alexandria for Athens) and the Alexandrians Ammonius and Asclepius appear to have acknowledged their beliefs quite openly, as later

---

38. Philoponus *in DA* 21,20-3.

did the Alexandrian Olympiodorus, despite the presence of Christian students in their classes.[39]

The teaching of Simplicius in Athens and that of the whole pagan Neoplatonist school there was stopped by the Christian Emperor Justinian in 529. This was the very year in which the Christian Philoponus in Alexandria issued his proofs of Creation against the earlier Athenian Neoplatonist Proclus. Archaeological evidence has been offered that, after their temporary stay in Ctesiphon (in present-day Iraq), the Athenian Neoplatonists did not return to their house in Athens, and further evidence has been offered that Simplicius went to Harrān (Carrhae), in present-day Turkey near the Iraq border.[40] Wherever he went, his commentaries are a treasurehouse of information about the preceding thousand years of Greek philosophy, information which he painstakingly recorded after the closure in Athens, and which would otherwise have been lost. He had every reason to feel bitter about Christianity, and in fact he sees it and Philoponus, its representative, as irreverent. They deny the divinity of the heavens and prefer the physical relics of dead martyrs.[41] His own commentaries by contrast culminate in devout prayers.

Two collections of articles by various hands have been published, to make the work of the commentators better known. The first is devoted to Philoponus;[42] the second is about the commentators in general, and goes into greater detail on some of the issues briefly mentioned here.[43]

39. For Simplicius, see I. Hadot, *Le Problème du Néoplatonisme Alexandrin: Hiéroclès et Simplicius* (Paris 1978); for Ammonius and Asclepius, Koenraad Verrycken, *God en wereld in de Wijsbegeerte van Ioannes Philoponus*, Ph.D. Diss. (Louvain 1985); for Olympiodorus, L.G. Westerink, *Anonymous Prolegomena to Platonic Philosophy* (Amsterdam 1962).

40. Alison Frantz, 'Pagan philosophers in Christian Athens', *Proceedings of the American Philosophical Society* 119 (1975), 29-38; M. Tardieu, 'Témoins orientaux du *Premier Alcibiade* à Harrān et à Nag 'Hammādi', *Journal Asiatique* 274 (1986); id., 'Les calendriers en usage à Harrān d'après les sources arabes et le commentaire de Simplicius à la *Physique* d'Aristote', in I. Hadot (ed.), *Simplicius, sa vie, son oeuvre, sa survie* (Berlin 1987), 40-57; id., *Coutumes nautiques mésopotamiennes chez Simplicius*, in preparation. The opposing view that Simplicius returned to Athens is most fully argued by Alan Cameron, 'The last day of the Academy at Athens', *Proceedings of the Cambridge Philological Society* 195, n.s. 15 (1969), 7-29.

41. Simplicius *in Cael.* 26,4-7; 70,16-18; 90,1-18; 370,29-371,4. See on his whole attitude Philippe Hoffmann, 'Simplicius' polemics', in Richard Sorabji (ed.), *Philoponus and the Rejection of Aristotelian Science* (London and Ithaca, N.Y. 1987).

42. Richard Sorabji (ed.), *Philoponus and the Rejection of Aristotelian Science* (London and Ithaca, N.Y. 1987).

43. Richard Sorabji (ed.), *Aristotle Transformed: the ancient commentators and their influence* (London and Ithaca, N.Y. 1990). The lists of texts and previous translations of the commentaries included in Wildberg, *Philoponus Against Aristotle on the Eternity of the World* (pp. 12ff.) are not included here. The list of translations should be augmented by: F.L.S. Bridgman, Heliodorus (?) in *Ethica Nicomachea*, London 1807.

I am grateful for comments to Henry Blumenthal, Victor Caston, I. Hadot, Paul Mercken, Alain Segonds, Robert Sharples, Robert Todd, L.G. Westerink and Christian Wildberg.

English-Greek Glossary

ability: *dunamis*
able: *dunasthai*
absence: *apousia*
accidental: *kata sumbebêkos*
account: *logos*
accurately: *akribôs*
act: *poiein*
active: *poiêtikos*
activity: *energeia*
actuality: *entelekheia* (see n. 116),
   energeia
addition: *prosthêkê*
affected: *paskhein*
affection: *pathêma, pathos*
age (v): *palaiousthai*
aggregate: *sunkrinein*
agreement: *akolouthos*
aid: *sunergon*
air: *aêr, pneuma*
airy: *pneumatôdês, pneumatikos*
alive (v): *zên*
alteration: *alloiôsis*
always: *aei*
amber: *êlektron*
analogous: *analogos*
animal: *zôion*
anomeomerous: *anomoiomerês*
antithesis: *antithesis*
apprehension: *antilêpsis*
around: *kuklos*
ashes: *tephra*
assimilate: *proskrinein*
attempt: *epikheirêsis*
art: *tekhnê*

bark: *phloios*
be: *einai*
become: *ginesthai*
become thin: *leptunesthai*
beginning: *arkhê*
belly: *gastêr, koilia*

birdlime: *ixos*
blood: *haima*
blow out (v): *apopnein*
bodily: *sômatikos*
body: *sôma*
boil (v): *exepsein, hepsein, zein*
boiled: *hepsêtos, hephthos*
boiling: *hepsêsis, hephthos*
bone: *ostoun*
both: *amphoteros*
both together: *sunamphoteros*
boundary: *horos*
bowel worm: *terêdôn*
bowl: *phialê*
break: *diaspasthai*
bring to perfection: *sunepitelein*
broken (v): *katessesthai*
bulk: *onkos*
burn: *kaiein, puroun*
burning: *purôsis*
burnt: *proskaien*

cast out: *ekballein*
catch fire: *ekpurousthai*
cause: *aition, aitios, aitia*
cave: *spêlaion*
chain: *halusis*
change: *metabolê*
change (v): *metaballein*
charcoal: *anthrax*
charcoal-yielding: *anthrakeutos*
chest: *thôrax*
chilling: *katapsuxis*
class: *genos*
clay: *pêlos*
cloud: *nephos*
cohesive (v): *sunaptesthai*
cold: *psuxis, psukhos, psukhros* (adj.)
coldness: *psukhrotês*
colour: *khrôma*
colour (v): *khrômatizein*

combination: *sumplokê*
combine: *sunkrinein*
combustible: *kaustos*
come to be: *ginesthai*
come to be flaming: *phlogousthai*
common: *koinos*
compacted: *sunistanai*
compare: *eikazein*
comparison: *sunkrisis*
complete: *teleios*
compose: *sunistanai*
composed: *sunkeisthai*
composition: *sunthesis, sustasis*
compound: *miktos, sunthetos*
compounded: *sunkeisthai*
comprehend: *antilambanesthai*
compressed: *sunthlibesthai*
compressible: *piestos* (see n. 217)
concave (v): *katakamptein*
concentrate: *sunagein*
concentrated (v): *athroizesthai* (see n. 142)
concoct: *pettein*
concoction: *pepsis*
condensed: *sunistanai* (see n. 65)
condensing: *sustasis*
conform: *skhêmatizein*
conformed: *suskhêmatizesthai*
congealed: *sunistanai*
connate: *sumphutos*
consider: *prokheirizesthai*
consistency: *sustasis*
constituted: *sunistanai*
constructed: *kataskeuê*
consume: *analiskein*
consumed: *sunanaliskein*
continuous: *sunekhês*
contract (v): *puknousthai, sustellesthai*
contracted: *sunienai*
contrariety: *enantiôsis*
contribute: *suntelein*
convex (v): *anakamptein*
cool (v): *psukhein*
cook: *opsopoios*
copper: *khalkos*
co-produce: *suntelein*
corpse: *nekros*
coupling: *sunduasmos*
craft: *tekhnikos*
craftsmanship: *dêmiourgia*

create: *apotelein*
criterion: *kritêrion*
crumble: *diapiptein*
curd: *turos*
curdle: *turousthai*

dead person: *nekros*
deficiency: *elleipsis, endeia*
deficient: *endeês*
deficient (v): *endein*
define: *horizein*
defining: *horistikos*
definition: *horismos, logos*
delicate: *leptos*
density: *puknotês*
depart: *aperkhesthai*
destroy: *phtheirein*
destruction: *phthora*
destructive agent: *phthartikon*
determine: *diorizein, horizein*
difference: *diaphora*
digestion: *katergasia*
directly: *prosekhôs*
discharge: *rheuma*
disintegrate: *diapiptein*
displaced: *antiperiistasthai*
disproportion: *asummetria*
dissipate: *diapnein, sunexatmizein*
dissolution: *diakhusis, lusis*
dissolve: *luein*
distinguish: *krinein*
divide: *diairesthai*
do: *poiein*
draw: *helkein*
drenched (v): *tengeisthai*
drier: *xêroteros*
drinking: *rhophêsis*
dross: *skôria*
drug: *pharmakon*
dry: *xêros*
dryness: *xêrotês*
dry out: *xêrainein*
dry up: *anaxêrainein*

earth: *gê*
earthen: *geêros*
earthy: *geôdês*
easily separated: *euapolutos*
easily shattered: *euthraustos*
eating: *edôdê*

ebony: *ebenos*
edible character: *edôdismos*
efflux: *katarrous*
eject: *ekkrinein*
ejection: *ekkrisis*
element: *stoikheion*
end: *telos*
equivocally: *hômonumôs*
escape: *ekluein*
especially: *proêgoumenôs*
essential: *kath' hauto*
evaporate: *diatmizesthai, exatmizein, exikmazein*
evenly: *homalôs*
evidence: *pistis*
excess: *huperbolê*
excrement: *apokrisis*
excreta: *perittôma*
excrete: *ekkrinein*
exhalation: *anathumiasis*
exist: *einai*
existence: *hupostasis*
exist within: *enuparkhein*
eye: *ophthalmos*

face: *prosôpon*
fall short: *endein*
fashion: *tropos*
fat: *pion*
fibre: *is*
fig-juice: *opos*
final: *telikos*
finer: *leptoteros*
fire: *pur*
first: *prôtos*
fissile: *skhistos*
fit (v): *epharmozein*
flame: *phlox*
flavour: *khumos*
flesh: *sarx*
flexible: *endidon, endotikos, kamptos*
float (v): *epipolazein*
flow: *rhusis*
flow off: *aporrein*
fluid: *hugros*
food: *opson, sition*
foreign: *allotrios*
form: *morphê*
formation: *sustasis*
fragility: *thrausis*

frankincense: *libanôtos*
fried (v): *tagênizesthai*
fried: *tagênon*
fruit: *karpos*
frying: *tagênisis*
fumable: *thumiatos*
fumable (v): *ekthumiasthai*
fume (v): *thumiasthai*
fumes: *kapnos, thumiasis*

garment: *himation*
gather: *sunagein*
gathered (v): *athroizesthai*
generally: *katholou*
generate out of: *epigênnan*
generated: *gennasthai*
generation: *genesis*
get rid of: *sunekkrinein*
give: *apodidonai*
give form: *eidopoiein*
gnat: *kônôps*
glass: *huelos*
glue: *kolla*
gluing: *kollêsis*
gold: *khrusos*
good odour: *euôdia*
growing: *epidosis*
growth: *auxêsis*
gum: *kommi*

hair: *thrix*
hand: *kheir*
happen: *sumbainein*
hard: *sklêros*
harden: *sklêrunein*
harder: *sklêroteros*
hardness: *sklêrotês*
have: *ekhein*
head: *kephalê*
heat (v): *thermainein*
heat: *thermê, thermotês*
heated: *puraktousthai*
heated previously: *prothermainesthai*
heavier: *baruteros*
help: *sunergein*
heterogeneous: *anomoiogenês*
hold: *ekhein*
hold together: *summenein, sunekhein*
homeomers: *homoiomerês*
homogeneous: *homogenês*

honey: *meli*
horn: *keras*
hot: *alea, thermos*
hotter: *thermoteros*

ice: *krustallos*
ignite: *exaptein*
imitate: *mimeisthai*
impact: *plêgê*
imperfection: *ateleia*
impossible: *adunatos*
impressed (v): *thlasthai*
impressible: *thlastos*
inability: *adunamia*
incapacity: *adunamia*
included: *sumperilambanein*
incombustible: *akaustos*
incomplete: *atelê*
inconcoction: *apepsia*
incorporate: *agein*
increase (v): *auxein*
indissoluble: *alutos*
induction: *epagôgê*
inedible: *abrôtos*
inhere: *huparkhein*
inherent (v): *enuparkhein*
inside: *bathos*
insoluble: *alutos*
instrument: *organon*
interconnect: *suntitrasthai*
interlock: *epallassein*
intermediate: *metaxu*
internal organs: *splankhnon*
intestinal worm: *helmins*
intoxicate: *methuskein*
iron: *sidêros*
irregular: *anômalos*

juice: *khumos*
juxtapose: *parakeisthai*

kettle: *khalkeios*
kind: *genos*
kindle: *zôpurein*

lack: *endeês, sterêsis*
lack (v): *ekleipein*
last: *eskhatos*
lay hold of: *antilambanesthai*
lead: *molibdos*

leather: *himas*
leave: *phullon*
lesser: *hêtton*
lid: *pôma*
lighten: *leukainesthai*
lightning: *astrapê*
like: *analogos*
likeness: *eikôn*
lime: *titanos*
limit: *horos*
liquefied humour: *suntêgma*
loosen: *aneinai*
lose: *apollunai*
lower: *katô*
lye: *konia*

make one: *henoein*
malleable: *elatos*
manner: *tropos*
marrow: *muelos*
master (v): *epikratein, kratein*
material: *hulikos*
mathematical: *mathêmatikos*
matter: *hulê*
mean: *mesos, mesotês*
median status: *mesos*
melt: *têkeisthai*
meltable: *têktos*
melting: *têxis*
mention: *mnêmoneuein*
mercury: *hudraguros*
metal: *metalleuton*
metaphor: *metaphora*
metaphorically: *metaphorikôs*
meteorology: *meteôrologikê*
milk: *gala*
mined: *metalleuesthai*
misshapen (v): *diastrephesthai*
mix: *mignunai*
mixture: *migma, mixis, miktos*
moderation: *summetros*
moist: *hugros*
moisten: *hugrainein*
moister: *hugroteros*
moistness: *hugrantikê*
moisture: *hugrotês*
more earthy: *geêroteros*
more formal: *eidikôteros*
more generic: *genikôteros*
more watery: *hudatôdesteros*

mosquito: *empis*
motion: *kinêsis*
moulding: *plasis*
move: *kinein*
move aside: *antimethistasthai*
mud: *pêlos*
mulas: *mulê*
mulia: *mulia*
must: *gleukos, glukus*
musty: *glukus*
myrrh: *smurna*

name: *onoma*
name (v): *onomazein*
narrower: *stenôteros*
natural: *phusikos*
naturally: *phusikôs*
nature: *phusis*
nearer: *enguterô*
nearest: *engutatô*
necessarily: *anankê*
nitron: *nitron*
noise: *psophos*
noiseless: *apsophos*
non-existent: *anuparktos*
non-fumable: *athumiatos*
non-impressible: *athlastos*
non-malleable: *anêlatos*
non-nutritive: *atrophos*
nourish: *trephein*
nutrition: *trophê*
nutritive: *threptikos*

odour: *osmê*
oily: *liparos*
old age: *gêras*
olive-like: *elaiôdês*
olive-oil: *elaios*
ooze: *exienai*
operation: *ergon*
opposite: *antikeimenos*
opposite(s): *enantios*
organise: *sunistanai* (see intro.)
outside: *exô*
overcome: *huperballein, katiskhuein, kratein*
overpower: *biazesthai*
overpowering: *epikrateia*

parallel: *parallêlos*

part: *meros, morion*
passive: *pathêtikos*
peculiar: *idios*
perceive: *aisthanesthai*
perceiving faculty: *aisthêtikos*
perceptible: *aisthêtos*
perfect (v): *teleioun*
perfection: *teleiôsis, teleiôtês*
pericarpia: *perikarpion*
person: *anthrôpos*
phlegm: *phlegma*
physician: *iatros*
piercing: *tmêtikos*
pipes: *aulos*
pitch: *pissa*
plant: *phuton*
plastic: *plastos*
pool: *limnê*
pore: *poros*
potential: *dunamis*
pottery: *keramos*
power: *dunamis*
preboil: *proepsein*
predisposition: *epitêdeiotês*
pre-exist: *proüparkhein*
presence: *parousia*
preserve: *phulassein, sôzein*
pressure: *ôsis*
prevent: *kôluein*
preventing: *kôlutikos*
primarily: *mallon*
primary: *prôtos*
principle: *arkhê*
produce: *ergazesthai, poiein*
production: *ergasia*
proper: *kurios, oikeios*
proportion: *analogia, summetria*
pulp: *sustasis* (see n. 68)
purification: *apokatharsis*
purify: *apokathairein*
pus: *puon*
putrefaction: *sêpsis*
putrefy: *sêpesthai*
pyrimachus: *purimakhos* (see n. 166)

quarried, *oruktos*
quick: *khutos*

ratio: *analogia*
rational: *logikos*

rather: *mallon*
raw: *ômos*
rawness: *ômotês*
reasonably: *eulogôs*
receive: *anadekhesthai*
reckoned: *sunarithmesthai*
remoistened (v): *anugrainesthai*
remove: *exairein*
rennet: *putia*
residue: *perittôma*
resist: *antitupoun*
resistant: *antibatikos*
rest: *loipos*
retain: *phulassein*
rheum: *lêmê*
ripe: *pepôn*
ripened: *pepainesthai*
ripening: *pepansis, pepsis*
roast: *optan*
roasted: *optos*
roasting: *optêsis*
rot: *sêpesthai*
rottenness: *saprotês*

salt: *hals*
saw: *priôn*
searing: *stateusis*
seasoning: *êdusma*
seed: *sperma*
seethe: *zein*
semen: *gonê*
sense: *aisthêsis*
sensible: *aisthêtos*
separate: *khôris*
separate (v): *apokrinein, khôrizein*
separate out: *diakrinein*
separated off from: *sunapokrinein*
serum: *ikhôr*
serum-like: *ikhôroeidês*
shape: *skhêma*
shattered (v): *thrauesthai*
show (v): *deiknunai*
shut: *muein*
shut inside: *enkatakleisthai*
sight: *opsis*
sign: *sêmeion*
silver: *arguros*
similar: *homoeidês*
similarity: *homoeidôs*
simmer: *molunein*

simmering: *molunsis*
simple: *haplous*
sinew: *neuron*
smelly steam: *knissa*
smoke: *kapnos*
smokeless: *akapnos*
smoky: *kapnôdês*
soft: *malakos*
soften: *malassein* or *malattein*
softer: *malakôteros*
softness: *malakotês*
solid: *stereos*
solidifiable: *pêkton*
solidify: *pêgnunai, pêssein*
solidity: *pêxis*
sooty: *lignus*
soul: *psukhê*
split: *skhizesthai*
splitting: *skhisis, tmêtos*
spoil: *diaphtheirein*
squashable: *pilêtos* (see n. 220)
squeezed (v): *thlibesthai*
squeeze out: *ekthlibein*
stag: *elaphos*
stalactite: *pôros* (see n. 184)
standing: *stadaios*
steel: *stomôma*
stickiness: *gliskhrotês*
sticky: *gliskhros*
stiff: *psathuros* (see n. 219)
strained through: *diêthesthai*
structure (v): *sunistanai*
structure (subst.): *sunistanai*
substance: *ousia*
substrate: *hupokeimenos*
suitable: *epitêdeios*
sum up: *sunkephalaiousthai*
sun: *hêlios*
surface: *epiphaneia*
surroundings: *periekhon*
sweet: *glukus*

take a shape: *skhêmatizein*
tangible: *haptos*
tear: *dakruon*
tensility: *tasis*
there: *entautha*
thick: *pakhus*
thicken: *pakhunein*
thickening: *pakhunsis*

thicker: *pakhuteros*
thickness: *pakhutêtos*
thinner: *leptoteros*
thunder: *brontê*
time: *khronos*
tin: *kassiteros*
tool: *organon*
touch: *haphê*
tractile: *helktos*
tractility: *helxis*
transfer: *metapherein*
transition: *metabasis*
tumour: *phuma*

ultimate: *eskhatos*
unboiled: *anepsêtos*
uncococted: *apeptos*
undetermined: *aoristos*
undrenchable: *atenktos*
undrinkable: *apotos*
unevenness: *anômalia*
unite: *sumphuein*
unmixed: *akratos, amiktos*
unslaked lime: *asbestos*
unsure (v): *aporein*
upper: *anô*
urine: *ouron*

vapour: *anathumiasis, atmis*
vaporise: *atmizein*
vaporous: *atmidôdês*
very cold: *psukhrotatos*

vessel: *angeion*
violent: *biaios*
vinegar: *oxos*
vomited: *emein*

water: *hudôr*
watery: *hudarês, hudatôdês*
wax: *kêros*
way: *tropos*
weaken: *exasthenein, katamarainein*
weaker: *asthenestera*
weakness: *asthenês*
well-springs: *pêgê*
whey: *orrôdês, orros*
why: *dia ti*
wind: *pneuma*
wine: *oinos*
winey: *oinôdês*
winged creature: *ptênos*
withering: *auansis*
without qualification: *haplôs*
without soul: *apsukhos*
withstand: *antibainein*
wood: *xulon*
wooden: *xulinos*
wool: *erion*
work: *ergon*
worm: *skôlêx*
wrought: *katergazesthai*

yield (v): *eikein*

Greek-English Index

This index was compiled by Ian Crystal. References are to the page and line numbers of the *CAG* edition, which appear in the margin of this translation.

anathumiasis, vapour, that which is vaporous, exhalation 204,7.20.21; 207,18; 213,15.16.18.20; 218,13.31.32

anaxêrainein, to dry up, 196,18.23; 197,6

aneinai, to loosen, 180,35

anêlatos, non-malleable, 216,30

anepsêtos, unboiled, 190,34

angeion, a vessel, 193,17; 196,8.9

anô, upper, up, 197,10.14.17.20; 208,23

anômalia, unevenness, 195,22.24.25; 206,28

anômalos, irregular, irregularly, 215,12

anômalôs, irregular, 215,10.11.15

anomoiogenês, heterogeneous thing, 180,26

anomoiomerês, anomeomerous, 213,38; 219,18.19.20.22; 223,7.12.13.21.26; 224,4.11; 225,2.3; 226,13.20.22; 227,17.21

anônumôteros, harder to name, 197,23

anthrakeutos, charcoal-yielding, 218,21

anthrax, charcoal, 218,17.21

anthrôpos, person, 188,28; 223,26.27.28.29; 224,1; 226,23

antibainein, to withstand, 200,1

antibatikos, resistant, 201,15

antikeimenos, opposite, 179,21; 181,27; 186,15.19.20.28.32; 188,14.16; 197,23

antilambanesthai, to comprehend, lay hold of, 200,32; 201,5.7.8

antilêpsis, apprehension, the apprehending of, 200,27

antilêptikos, being capable of apprehending, 200,25; 201,1

antilêptos, apprehended, 201,4

antimethistasthai, to move aside, 216,23

antiperiistasthai, to be displaced, 200,2.4.5; 202,32

antithesis, antithesis, 214,6.7.8; 219,6.9

antitupoun, to resist, 200,1

anugrainesthai, to be remoistened,

to become moist again, 206,23; 207,4

anuparktos, non-existent, 180,1

aoristos, undetermined, 180,15.18; 183,15; 188,5; 189,29; 191,20.23.26.29; 192.19; 193,12; 194,3.24.35; 195,1; 198,27; 199,9; 200,17.18.19

apêktos, not-solidifiable, cannot be solidified, 214,8.34; 215,2

apepsia, inconcoction, 182,19.21.24.28.29; 185,32; 186,1.4.5.8; 188,10.12.15.18; 189,27.28; 191,13; 194,30; 195,1.8.11.27.31; 197,22.23.29.31.32

apeptos, uncococted, 186,24; 188,7; 190,3; 197,9; 209,34

aperkhesthai, to depart, 210,1.3

apilêtos, can be stamped, 217,10

apiôn, a departing, 205,27

apo, after, from, 188,33; 190,23; 191,21.30 et passim

apodidonai, to give, 181,6; 184,14; 201,15; 211,27; 222,3; 226,31.33; 227,4

apokathairein, to purify, 207,21.23

apokatharsis, purification, 207,20

apokrinein, to separate, 185,9; 209,34

apokrisis, excrement, 197,11

apollunai, to lose, 222,23

apopnein, to blow out, 204,9

aporein, to be unsure, 220,4.9

aporôtatos, a serious difficulty, 208,17

aporrein, to flow off, 207,25.29

apotelein, to make, create, 188,8.27; 198,32

apotos, undrinkable, 190,19

apousia, loss, absence, 207,24; 214,18.20

apsophos, noiseless, 219,11

apsukhos, without soul, 225,19

arkhê, beginning, principle, 183,9.24; 186,35; 187,6.31; 195,3; 198,10.11.24; 212,23; 217,28; 227,10

arguros, silver, 180,25; 193,4; 213,13; 215,4; 220,29; 225,21; 226,16.22

Aristotelês, Aristotle, 179,2.3.4

asbestos, unslaked lime, 207,30

askhistos, that which is non-fissile, 216,33

asthenês, weakness, 181,30; 182,17.18.25.28.29; 184,31; 217,12

asthenestera, weaker, 182,18; 184,31; 217,13

astrapê, lightning, 202,31

asummetria, disproportion, 182,17; 190,1.14.15

atêktos, not meltable, unable to melt, 212,15.16; 214,9.32; 215,21.24; 221,1

atelê, incomplete, 197,25

ateleia, imperfection, 188,10.12.15; 189,34; 190,13

atenktos, incapable of being drenched, 215,22.23.25.26

athlastos, non-impressible, 216,19.21.22

athroizesthai, to be gathered, to be concentrated (see n. 142), to be collected, to be rounded up, 202,30; 203,1; 210,5.6

athumiatos, non-fumable, 217,14

atmidôdês, vaporous, 204,20.21; 213,16

atmis, vapour, 217,18; 218,2.3.5

atmizein, to give off vapour, vaporise, 206,24; 209,21.22; 210,21

atrophos, non-nutritive, 211,10

auansis, withering, 183,2

aulos, pipes, 224,6.7

auxein, to increase, 182,27

auxêsis, growth, 181,26

bathos, middle, inside, 196,25.28; 200,6.15

baruteros, heavier, 194,5.16

biaios, violent, 181,31; 183,4

biazesthai, to overpower, 206,33

brekhesthai, to get wet, 203,13; 215,27

brontê, thunder, 202,31

dakruon, tear, 220,22.25

deiknunai, to show, 179,8; 181,4.13; 183,21; 184,6.18; 188,21; 189,12; 191,13.36; 196,3; 198,20; 199,27; 203,28; 205,4.21.24; 210,35; 211,10; 221,21; 223,26; 224,6; 226,1

dein, necessarily, must, 182,9; 190,6; 199,21; 212,6

dektikos, can receive, capable of receiving, 201,10; 217,12

dêlôtikos, indicates, makes clear, 182,24; 199,14

dêmiourgesthai, to be fabricated, 219,30; 221,31

dêmiourgia, craftsmanship, 222,2

dia, for, because, through, 179,19; 180,9.10.31 et passim

diairesthai, to divide, 200,3

diakhusis, dissolution, 202,1

diakrinein, to separate out, 180,23; 181,10; 194,4.8.13; 221,21; 225,27

diakritikos, capable of being separated, 181,8

dialuesthai, to be loosened, 203,21

diapherein, to be different, 179,19; 181,8; 186,3; 194,20; 196,3.5; 198,19; 206,6; 213,24.28.33; 214,6; 217,1.25; 219,7.8; 226,13.15.20

diaphora, difference, 181,17.18; 192,31; 213,25.26; 214,5; 216,18.27; 219,9.13; 226,10

diaphtheirein, to spoil, 181,32; 182,23; 211,23

diapiptein, to disintegrate, to crumble, 183,16; 199,2

diapnein, to dissipate, 184,15

diaspasthai, to break, 217,9

diastrephesthai, to become misshapen, 206,26.27

diatmizesthai, to evaporate, 190,9

didonai, to give, 212,1

diêthesthai, to be strained through, 221,27

dio, hence, for, 183,19; 184,29; 185,2; 186,23; 188,2; 189,30; 190,6; 191,36 et passim

diorizein, to determine, determined thing (see *horizein*), 179,12; 180,22; 195,25; 222,28; 226,9

diugrainesthai, to be soaked through, 196,28

dokein, to seem, 187,3; 188,30; 191,32; 194,5; 201,29; 202,23.34; 213,4; 220,15

dunamis, ability, power, potential, 179,7.8.9.10; 181,20.21.26.28.33;

209,24.25.29.30;
210,2.7.22.23.26.29.34.35;
211,2.9.11.17;
212,10.13.14.15.28.33;
213,9.15.17.19.36.37; 214,29.36;
215,7.8.13.24.28.33; 216,3.8;
217,23; 218,22.26.27;
219,5.24.26.27.28.34.35.36;
220,2.4.9.10.12.15.32;
221,7.8.9.18.22.30;
222,6.7.8.16.20.24.32; 223,10.17;
224,31; 225,23

geêros, earthen, 189,18

geêroteros, more earthy, 189,19

genesis, generation, 179,6.13; 180,7;
181,19.20.22.23.24.27.34; 182,
3.10.15.30.31; 183,12;
185,12.13.16.17.26.27.32; 186,23;
198,7; 199,15; 207,17; 223,11; 227,7

genikôteros, more generic, 193,26

gennasthai, to be generated,
197,11.13.16; 209,11; 226,23

genos, kind, class, 188,20; 191,12;
223,9; 226,27.31

geôdês, earthy, 190,6.11; 199,19;
205,29; 207,21; 211,1

gêras, old age, 181,30; 183,2

ginesthai, to come to be, become,
occur, produce, cause, 179,9.11;
181,21.26.28.29.31;
182,3.11.13.19.25; 183,1.9.20.27
et passim

gleukos, must, 187,16.22.23;
193,7.14; 209,33

gliskhros, sticky, 203,19

gliskhrotês, stickiness, 203,18

glukus, must, musty, sweet, 214,36;
218,7; 220,13

gonê, semen, 222,22

haima, blood, 210,25.27.30;
211,8.10.14.15.17.18.22.23.25;
222,21; 223,30; 225,14

hals, salt, 208,5.14; 210,28.31; 214,29;
215,27

halusis, chain, 217,7

haphê, touch, 200,22.23.24.26.29;
201,1.6.7.8

haplôs, without qualification, 194,5;

200,8.10.17.18.19.22; 201,3.5.6.7;
209,30

haplous, simple, 181,19.22.24;
182,1.2.30.31; 199,10; 219,30;
224,11.19.30; 225,17

haptos, tangible, 179,7.16.17.18

hautos, itself, themselves, 181,1;
182,2; 183,30 et passim; *kath'
hauto*, essential 187,32

hêlios, sun, 207,7; 209,16; 221,30

helkein, to draw, 184,20

helktos, tractile, 215,15; 216,28

helmins, intestinal worm, 197,15

helxis, tractility, 226,7.14

heneka, for the sake of, 223,23;
224,30; 225,1.4

henoein, to make one, 180,22.27

hephthos, boiled, boiling, 190,34;
191,21.23; 192,11.15.16; 195,24;
196,17.26; 205,5; 220,26

hepomenôs, as a consequence, 216,8

hepsein, to boil, 182,20.21; 187,21;
191,9.17.18.20.22.34.35;
192,1.7.11.23.24.26.28.32;
193,1.4.6.9.19.21.22.28.30.31.33;
194,6.7.12.14.15.23.24.26.34;
195,14.15.18.19.21.23.25;
196,14.18.19.20.28.30; 197,3;
204,6.7; 209,20.22.26.33; 210,36;
211,7; 218,1; 220,8

hepsêsis, a boiling, 180,20; 186,3.6.8;
191,14.15.22.25.27.31; 192,18.21;
193,7.8.16.25.26.33.34;
194,2.8.18.19.20.24.28.29.30.31.33.
34; 195,27.32; 196,3.6; 197,5.21.33;
198,2; 211,3

hepsêtos, boiled, 192,20.34; 194,1

heteros, other, another, either,
188,27, 197,17; 208,19; 209,16 et
passim

hêtton, lesser, 184,19.20.29.33;
200,14.17; 213,27; 224,8.11

himas, leather, 215,16

himation, garment, 203,12.29

homalôs, evenly, 196,31

homalunein, to bring to a uniform
temperature, 195,20.22

homoeidês, similar, 190,23

homoeidôs, similarity, 188,32; 189,11

keramos, pottery, 190,29; 191,1.4.10;
206,16.24; 208,1; 211,28.30;
212,14.16; 214,23.28.29.33;
215,11.13; 216,22; 220,21.31
keras, horn, 207,9.10; 211,29; 215,19
kêros, wax, 204,26; 216,21
khalkeios, kettle, 210,5
khalkos, copper, 180,25; 193,4;
213,13; 215,24.25; 216,21; 225,20
kheir, hand, 224,1.2.7.10; 225,2;
226,14
khloros, green, 220,3
khôris, without, separate, 201,24;
224,29
khôrizein, to separate, 183,23;
203,21.30; 206,7.9.11.17; 207,21;
210,36; 211,5.6.13.20; 212,19
khrôma, colour, 200,25.26.34; 207,30;
208,36; 213,26; 219,10; 225,29
khrômatizein, to colour, 217,31
khronos, time, 208,34; 209,2.6
khrusos, gold, 180,24; 192,29.31.32;
193,2.9; 213,13; 220,28; 225,21;
226,16
khumos, flavour, juice, 190,18;
193,10; 199,20; 213,26; 218,10;
219,10
khutos, quick, 215,4
kinein, to move, 195,19.21; 197,28
kinêsis, moving, motion, 184,31;
195,7.9; 226,5.6.11.18; 227,10
kinêtikos, moving, 183,24; 187,31
knissa, smelly steam, 218,1.33
koilia, belly, 197,10.12.14.17.18.20
koinos, common, 182,30; 184,12;
185,31; 186,10; 189,1; 197,32;
210,22; 217,29; 219,34; 220,4.11;
222,16.20
koinôs, common, 214,10
koinoteros, more general, 185,18;
193,25; 195,32
kolla, glue, 199,1.5
kollêsis, gluing, 199,6
kôluein, to prevent, 196,33
kôlutikos, preventing, 209,17
kommi, gum, 220,23
konia, lye, 210,20
kônôps, gnat, 185,9
kratein, to master, to overcome,
182,16.17.20.22.32.33;

183,12.17.27.29;
184,20.24.25.27.28; 185,1.3.16.21;
187,28; 188,1.4; 189,10.15;
190,2.19.28; 191,3.5.8.11;
192,11.14.21.25.32.34; 195,16;
196,10.11; 197,8; 211,25
krinein, to distinguish, 219,12
kritêrion, criterion, 200,21; 219,14
krustallos, ice, 205,15; 214,16.30;
215,7
kuklos, around, 193,11
kurios, proper, 186,9; 188,30; 190,21
kuriôs, in the proper sense, 181,22;
186,26; 189,29; 191,25; 192,29;
193,8.34; 194,2.33; 195,29; 218,9;
223,27; 225,11.15.26

lêmê, rheum, 187,25; 189,25
leptos, delicate, 189,20; 190,18;
201,13.14
leptoteros, thinner, finer, 194,8;
212,21; 214,27
leptunesthai, to become thin,
194,9.10
leukainesthai, to lighten, 208,34.5;
209,2.3
libanôtos, frankincense, 218,31;
220,23
lignus, sooty, 217,33; 218,32
limnê, pool, 185,6
liparos, oily (bodies), 218,1.2.33;
219,2.3
logikos, rational, 223,30
logos, account, ratio, definition,
180,9; 181,5.11; 182,7.9; 184,6;
185,18; 186,12; 191,13; 194,28;
198,9; 200,11.13; 201,24; 204,14;
214,9; 219,12;
223,9.18.19.24.25.28.29;
224,16.22.24.25; 225,24;
226,3.25.26.33; 227,1
loipos, that which remains, the rest,
184,15; 185,24.26; 194,10; 211,20;
224,12
luein, to dissolve, 180,29;
204,28.30.35;
205,1.11.14.17.20.21.23; 206,23.25;
207,1.2.4.7.29; 211,30.31.32.34;
212,1.2.3.10.20.22; 214,19.20.24.
29.31; 215,31; 216,11; 220,9

lusis, a dissolving, dissolution, 205,18; 212,29

lutikos, capable of dissolving, 205,17; 208,9

lutos, dissolved, soluble, 204,30; 208,1.5; 211,27; 212,6; 214,32; 215,21

malakos, soft, 199,30.32; 200,1.4.7.8.9.10.12.16.18.19.22; 201,5.7.13.16.17.19.21.22; 202,9; 206,12.13.14.20; 217,5; 219,31.32.33

malakôteros, softer, 200,15; 201,7; 206,25.26.27; 215,23.28; 216,9.10

malakotês, softness, 199,29; 226,8.15

malaktos, softened, capable of being softened, can be softened, 206,13; 207,8; 211,29; 212,11; 215,5.6.8.16.18; 216,20; 221,7

malassein or *malattein*, to soften, 180,30.35; 207,9; 215,18.20

mallon, rather, primarily, 179,5; 183,31; 184,23.34; 220,32; 221,3 et passim

mathêmatikos, mathematical, 179,18

meli, honey, 180,19; 199,19; 205,5; 210,23; 214,36; 220,12.26

meros, part, 181,33; 182,19.24.25.29.30; 183,23; 185,5; 208,32; 223,7; 225,24

mesos, mean, median status, 201,10.33

mesotês, mean, 200,23.30.32; 201,4.8.9.11

metaballein, to change, 180,11.32; 182,4.14

metabasis, transition, 218,34

metablêtikos, able to bring about change, 180, 27

metabolê, change, 181,19.21.24.25.26; 182,3.10.11; 183,20; 185,14.17; 186,27.28; 188,17; 189,18.24; 193,14; 204,16; 207,20; 209,13; 210,15; 218,35

metalleuesthai, to be mined, 193,5; 214,31; 219,15; (as a substantive) metal, 213,12; 214,31

metalleuton, metal, 207,8; 213,19.21

metapherein, to transfer, 186,11

metaphora, metaphor, 188,33; 189,3; 190,23

metaphorikôs, metaphorically, 189,3.11; 192,29.30.35

metaxu, intermediate, 224,22

Meteôrologikê, The Meteorology (of Aristotle), 179,2.3; 213,22

meteôrologikê, meteorology, 179,4

methuskein, to intoxicate, 218,10.11

migma, mixture, 208,30; 220,29

mignunai, mix, 181,3; 189,31; 190,7; 193,3; 197,5; 198,13.25; 203,20; 206,2; 209,24; 210,1.9; 219,30

miktos, compound, composed, mixed, mixture, 198,15; 199,16; 203,6; 205,32; 206,12.17.19; 208,11; 212,33; 213,2.8; 220,6; 226,29

mimeisthai, to imitate, 197,3

mixis, mixture, 183,10.15; 198,26.33.35; 199,1.4.7.12.17; 208,36

mnêmoneuein, to mention, 197,31; 198,1; 199,5; 211,4

molibdos, lead, 213,14; 219,16

molunein, simmer, 182,21; 195,23; 217,31

molunsis, simmering, 182,19.20.23.28; 186,4.7.8; 191,14; 194,28.32.34; 195,1.7.11.21.27.29.31; 197,21; 198,2

morion, part, 180,25; 182,26; 183,6; 197,5; 225,14; 226,2.3.24; 227,17.19.21

morphê, form, 187,18.19

muein, to shut, 197,1

muelos, marrow, 222,22

mulê, mulas, 207,27; 208,3.4

mulia, mulia, 207,28; 208,3.4

nekros, a dead person, corpse, 223,26.27.29.31; 224,1.2; 225,30

nephos, cloud, 210,15

neuron, sinew, 210,31; 213,10; 215,16; 223,6; 225,23

nitron, nitron, 208,5.14; 210,29.31; 211,28; 214,14.19.29; 215,9.27.34; 216,1.10

oikeios, proper, its own,

180,15.16.18.20;
183,10.20.24.33.35;
184,2.9.12.15.17; 187,1;
188,1.5.11.12.32; 190,28; 191,2.3;
192,31; 193,10; 196,5; 198,27;
200,21.25; 203,11.14; 205,29;
213,28; 222,16; 224,33; 225,19
oinôdês, winey, 218,10
oinos, wine, 187,24; 190,10; 199,19;
203,10; 208,7; 209,28.29.32.34;
210,20; 218,7.9.10.11; 220,5.7.9.13;
221,26.29
ômos, raw, 190,29.32.34;
191,1.4.6.7.8.10; 192,10
ômotês, rawness, 186,4.6.8;
189,26.27.29.30.34; 190,12.24.25;
191,12; 197,33
onkos, bulk, 214,22; 215,29
onoma, name, 186,9.10; 188,30.33;
189,2.4.11; 190,21; 191,25; 192,31;
218,9; 225,10
onomazein, to name, 182,5; 188,34;
189,2.3.4
ophthalmos, eye, 187,25; 224,27.28;
225,8.9.10
opos, fig-juice, 211,3.5; 222,22
opsis, sight, 200,24.33
opson, food, 198,31.32
opsopoios, cook, 194,21
optan, to roast, 187,20; 191,36;
192,1.14; 196,7.10
optêsis, roasting, 186,3.7.8; 193,27;
196,1.3.6.11.12.13; 197,7.22.25.30;
198,1.3
optos, roasted, 192,11.15.16.17;
196,17.23; 197,24.26.27.28;
206,15.16.20.24.28
organon, tool, instruments, 187,2;
194,20.23; 223,22; 224,7
orrôdês, whey, 194,10
orros, whey, 193,13; 203,10; 206,17;
208,7; 210,20.35; 211,5.6.21
oruktos, quarried, 213,20; 219,17
ôsis, pressure, 182,12; 191,12; 216,26;
222,26
osmê, odour, 219,20
ostoun, bone, 183,6; 213,10; 223,6;
224,9
oudeteros, not either, neither,
208,19; 209,16

ouketi, not further, never, but not,
195,28; 196,24; 199,21; 204,35;
209,3 et passim
ouron, urine, 188,3; 190,26; 194,14;
203,10; 208,7; 210,20; 221,26.27
ousia, being, substance, 181,11;
185,30; 187,10.11; 190,9.11; 198,8;
203,13.16; 223,17; 224,20.24;
225,4.25; 226,1
oxos, vinegar, 208,8; 210,20

pakhunein, to thicken, 180,12.19.35;
189,17.19.33; 190,4.6.8.10.33;
194,4.9.14 et passim
pakhunsis, thickening, 194,4; 205,26
pakhus, thick, 201,15
pakhuteros, thicker, 180,20; 188,6.7;
189,20; 194,8; 206,21; 208,30;
209,5; 223,1
pakhutêtos, thickness, 218,14
palaiousthai, to age, 217,22
parakeisthai, to juxtapose, 180,27
parallax, not to meet, 216,4.7
parallêlos, the same way, parallel,
182,23; 193,25
parousia, presence, 202,6; 207,16;
214,19
paskhein, to undergo, be affected,
179,10.33.34; 181,2.12.18;
182,5.6.8.13.14.17.32; 183,16.27 et
passim; (as a substantive) the
subject, the patient, 202,6 et passim
pathêma, affection, 199,25
pathêtikos, passive, 179,10; 180,6.8;
181,5.14.16.18; 182,11.14.16;
185,15.21; 186,15.19.30.32;
188,14.16; 192,27; 198,4.9; 199,22;
202,15.19.20.23.25; 203,6;
213,24.33.34; 222,5; 225,22; 226,9
pathos, affection, 180,34; 181,2.18;
184,12; 186,20; 188,16; 197,24;
198,11.12.13.15.16.18.24; 199,29;
200,27.28.30; 201,20.25.27;
202,6.9.11; 204,15; 209,6.8;
213,29.31.33; 214,1.4; 216,18;
219,8; 221,11.15.16.21
pêgê, well-springs, 202,34
pêgnunai, to solidify,
180,12.17.27.35; 181,1; 194,12;
202,2.8.13 et passim

plastos, plastic, 216,24.25

plêgê, impact, 216,27

pneuma, pneuma, air, wind (see n. 226), 189,18.31; 204,19.20; 208,26; 217,19.26.27; 218,29

pneumatikos, airy, airy things, 189,16.17.30.32.33

pneumatôdês, airy, 191,27.30; 193,13; 195,5

poiein, to do, act, produce, 179,10; 181,10.11; 182,8.12.17.25.29.31; 183,17 et passim; (as a substantive) the agent, 182,28 et passim

poiêtikos, active, activities, active power, agent, 179.9; 180,6.8; 181,5.14.16.17; 182,4.11.15.16.33; 183,11.13.26.28; 184,7; 185,14.21; 198,6; 202,4.12.24.26.28; 205,3; 213,2.5.7.24.28; 219,21.25; 220,22; 221,19; 225,22; 227,10.14

pôma, lid, 210,5

poros, pore, 197,1; 214,21.27; 215,13.15.29.32; 216,1.4.6.8; 217,11.12

pôros, stalactite (see n. 184), 210,33; 221,2

pothen, source, 227,10

priôn, saw, 225,11,13; 226,19,21

Problêmata, The Problems (see n. 111), 197,18

proêgoumenôs, especially, chiefly, 183,13; 186,2; 219,12

proepsein, to preboil, 192,17

prokheirizesthai, to consider, 221,20; 226,30

prosekhôs, directly, 203,1

proskaien, to be burnt, 196,20; 197,8

proskrinein, to assimilate, 182,26; 186,26

prosôpon, face, 224,10.26; 225,2

prosthêkê, addition, 181,23

prothermainesthai, to be heated previously, to be heated first, 211,35; 220,16

prôtos, first, primary, 179,15.16.17; 181,13.19; 183,8.19; 186,13.21.22; 188,19; 191,7.15; 194,30.31.33; 195,28; 198,12.24; 199,25; 202,9.12; 203,5; 204,24; 205,20;

206,12.15.16.17.21.23.24; 207,3; 212,4; 213,37; 214,2.8; 222,14; 223,13; 224,15.19.30; 226,31; 227,2.20

proüparkhein, to pre-exist, 207,15

psathuros, stiff (see n. 219), 217,9

psophêtikos, capable of making a sound, 213,27; 219,11

psophos, noise, 213,26; 219,13

psukhê, soul, 186,16

psukhein, to cool, 202,30; 203,24.25; 204,2.9.12; 209,4.14; 210,14; 211,11.37; 220,16

psukhos, cold, 184,19; 202,5; 203,25.32; 207,4; 208,27.32; 209,17; 210,23; 218,8

psukhros, cold (adj.), 180,5.17.26.28.30.32; 181,3.7.9.14; 184,8.10.11.23; 185,2; 186,1; 189,9; 190,19; 195,6.9.14; 197,8; 202,4.7.22.23.24.28.30.32.34; 203,3; 204,28.31.35; 205,2.5.6.7.8.10.25.31; 206,1.8.22.30; 207,32; 208,5.9.10.19.20.25.26; 209,12.13.31.32; 210,10.11.17.18.19.21; 211,12.13.16.31.32.33.35; 212,1.2.5.9.29.31.35; 213,6; 214,12.14.17; 219,25.31; 220,10.11.14.15.18; 221,2.12.13.19; 222,4.7.28; 226,5.7.11; 227,15

psukhrotatos, very cold, 222,33

psukhrotês, cold, coldness, 179,7.15; 180,3.6.11; 182,4; 183,12; 184,12; 185,3.4; 188,12; 189,6.7.9; 195,3.13; 202,21.27; 213,3.4; 222,4.31.34; 226,4.17

psuxis, cold, 204,4; 207,5; 208,28; 220,23.24.27.32

ptênos, a winged creature, 199,23

puknoteros, too dense, 212,20

puknotês, density, 192,23; 214,25

puknousthai, to contract, 196,25.32; 202,29; 204,16; 211,37

puon, pus, 187,24; 189,25

pur, fire, 180,2.19.22.24.31; 182,20; 183,26.28.29.30.32; 191,10.30.36; 192,2.14.28; 193,1.3.6.11.17; 196,7.8.16.18.19.20.24.26;

197,26.28; 199,21; 202,19;
204,7.32.35;
205,1.11.14.16.21.22.23.24.30.31.32;
207,19.25; 208,4.12.15.16.34;
209,1.6.20.31; 210,6.21;
212,8.17.22; 214,24; 215,18.26;
217,12.13.20.25; 218,28.35; 220,17;
221,6.8.14.32; 222,30; 224,12;
225,22
puraktousthai, to be heated, 192,5
purimakhos, pyrimachus (see n.
166), 207,25
purôsis, a burning, 222,12
puroun, to burn, 192,5; 217,15.18.29;
223,1.2
putia, rennet, 194,11

rheuma, discharge, 187,25
rhophêsis, drinking, 193,23.32
rhusis, flow, 210,33; 217,27

saprotês, rottenness, 183,4
sarx, flesh, 183,5; 191,22; 210,31;
213,10; 223,5; 224,9.32.34;
225,13.23
sêmeion, sign, 183,7; 188,2.4;
192,4.10; 193,30; 204,28; 208,22;
209,30; 210,4; 211,17.21
sêpesthai, to rot, putrefy,
183,5.22.25.27.28 et passim
sêpsis, putrefaction, 182,30;
183,1.3.4.7.8.19.23.27.32.34;
184,5.6.7.8.11.12; 185,22.23; 195,4;
197,13.19; 222,14.19
sidêros, iron, 192,6;
207,9.11.14.20.23; 212,4.7; 213,13;
215,18; 226,17,21
sition, food, 221,28
skhêma, shape,
180,13.14.15.16.18.20; 183,10;
198,29; 224,6; 225,29.30.32
skhêmatizein, to take a shape,
conform, 180,14.21; 198,28.29
skhisis, a splitting, 217,1
skhistos, fissile, capable of being
split, 216,33; 217,1.2
skhizesthai, to split, 216,33
sklêros, hard, 181,1; 199,30.31.32;
200,7.8.9.10.12.16.18.19.21.22.34;
201,1.2.3.6.16.17.19.21.22; 202,9;

207,26; 216,22; 219,31.32; 220,14;
221,13
sklêroteros, harder, 180,29;
195,22.24; 200,13.14; 201,8; 206,22;
215,29.30; 216,9.21
sklêrotês, hardness, 199,29; 226,8.15
sklêrunein, to harden, 180,29
skôlêx, worm, 185,9
skôria, dross, 207,21
smurna, myrrh, 220,23
Sôkratês, Socrates, 225,12
sôma, body, 179,15.18; 181,13; 182,1;
185,11.30; 192,19.26.33; 194,34;
195,1; 198,11.13.14.17.19.24;
199,8.10.12.13.15.25.27.31;
200,7.28.29; 201,16.19.20.22.23;
202,10.11.18.25; 203,6; 212,34;
213,1.3.5.6.8.10.32.35.37;
214,2.5.6.11.32; 216,12.26.27.29;
217,5.16; 218,15.23; 219,7.8.15.29;
222,4.6.8.20.25; 223,4.23;
224,11.19.30; 225,17.20.21.29.30;
226,1
sômatikos, bodily, 199,25; 201,25
sôzein, to preserve, 194,18; 201,13;
215,32; 222,21; 224,33; 225,29
spêlaion, cave, 221,2
sperma, seed, 188,26; 219,21
splankhnon, internal organs, 225,1
stadaios, standing, 184,29
stateusis, searing, 186,4.7.8;
197,24.29; 198,2
stenôteros, narrower, 214,22
stereos, solid, 204,18; 210,30; 211,18
stereotatos, extremely solid, 222,35
sterêsis, lack, being deprived, 214,30;
220,17.18.19; 222,35
stoikheion, element,
179,9.12.15.17.19; 180,1; 182,1;
223,13; 224,12.15.29
stomôma, steel, 207,22
sumbainein, to happen, 187,26;
195,17; 196,17; 211,36; 221,17;
222,32; *kata sumbebêkos*,
accidental, 187,32; 202,28
summenein, to hold together, 183, 34
summetria, proportion, 182,9
summetros, moderation, 201,12
sumperilambanein, to be included,
185,20

sumphuein, to unite,
180,11.12.23.25.32 et passim
sumphutos, naturally, connate,
194,27; 203,30
sumplokê, combination, 179,20
sunagein, to concentrate, to gather,
180,12; 202,29; 220,25
sunamphoteros, two, both together,
183,11; 219,2; 221,18; 224,17;
227,12
sunanaliskein, to be consumed,
217,21
sunapokrinein, to be separated off
from, 185,10
sunaptesthai, to be cohesive, 183,15
sunarithmesthai, to be reckoned,
201,28
sunduasmos, coupling, 179,8
sunekhein, to hold together, 222,21
sunekhês, continuous, 217,27
sunekkrinein, to get rid of, 206,31;
220,28
sunepitelein, to bring to perfection,
187,3
sunergein, to help, 187,4
sunergon, aid, 187,4
sunexatmizein, to dissipate, 184,16;
203,32; 204,4; 206,10; 207,5
sunienai, to be contracted, 196,32;
217,9
sunistanai, to organise (see intro.),
to structure, to be constituted, to be
condensed (see n. 65), to be
congealed, to compose, to be
compacted, 183,4; 185,10.25.30;
188,1; 189,17; 190,5.28.30.32;
194,11; 195,6; 198,8; 199,2.7;
201,22; 204,16.19.22; 205,28;
208,28; 210,5.32; 211,1.3.22;
213,9.11.13.23.35.36; 219,22;
220,2.8.13; 226,12.20.28; 227,18;
(as a substantive) structure, 187,24
sunkeisthai, to be compounded,
composed, 199,29; 206,29; 223,22;
224,26
sunkephalaiousthai, to sum up,
194,1
sunkrinein, to combine, aggregate,
180,23; 181,10; 220,20
sunkrisis, comparison, 200,11

sunkritikos, capable of aggregating,
181,7.9
suntêgma, a liquefied humour, 222,19
suntelein, to contribute, to
co-produce, 189,8; 221,4.6
sunthesis, composition, a
compounding, 199,27; 213,38
sunthetos, compound, 183,3; 200,7;
201,23; 202,10; 213,37; 214,1.4;
224,16.21
sunthlibesthai, to be compressed,
211,37
suntitrasthai, to interconnect,
216,2.5
suskhêmatizesthai, to be conformed,
180,17; 198,28.29
sustasis, consistency, pulp (see n. 68),
formation, condensing,
composition, 190,18; 192,26; 199,3;
210,15; 212,28.34; 213,1.2.14;
219,18.25.27; 221,19; 222,18; 226,6;
227,13
sustellesthai, to contract, 200,3.6.16

tagênisis, frying, 196,11
tagênizesthai, to be fried,
191,32.33.34.35; 192,1.2.5.13; 196,9
tagênon, fried, 191,33
tasis, tensility, 226,7.14
têkeisthai, to melt, 198,21; 204,18;
207,1.10.13.14.24.27.30; 208,4.6.8;
212,8; 214,20.23.31; 215,34;
216,6.7.9.10; 221,8.15
tekhnê, art, 196,30; 197,2.3; 226,22
tekhnikos, craft, 194,20.23
têktos, meltable, melted, 198,21;
212,11; 213,31; 214,9;
215,21.24.25.26.31; 218,23.24.25.26
teleioun, to perfect, 186,18.29.34;
215,14; 217,23
teleios, complete, 188,28; 224,3.4
teleiôsis, perfection, 186,14.16.31.35;
187,7; 188,24.25; 189,5.34
teleiôtatos, most complete, 227,4
teleiôtês, perfection, 188,13
telikos, final, 227,8
telos, end, 183,3; 187,7.8.10.12.14.17;
189,23; 193,19.21.23.28.29.30;
201,29.30; 213,21

tengeisthai, to be drenched,
215,33.34; 216,7.9
tenktos, capable of being drenched,
215,22.24.28
tephra, ashes, 221,27.32; 222,10.11.31
terêdôn, bowel worm, 197,15
tessares, four, 179,7.14.19.20; 180,5;
181,13.33; 222,5
têxis, melting, 204,22; 212,24.27.29.32
thalassa, sea, 185,5
theios, divine, 199,16
thermainein, to heat, 187,21;
190,5.8.9.10; 191,32; 192,5.9;
193,1.12; 194,3.16; 196,26.31.32;
202,5; 203,23; 204,2.3.12; 206,7.23;
207,4.10; 215,18.19; 217,16.20;
218,6; 220,30
thermê, heat, 184,12; 195,10; 221,27
thermos, hot, 180,4.18.28.29.32;
181,3.6.7.14; 182,8;
183,13.25.26.29.31.32;
184,6.8.10.11.15.17.20.21.22.23.24.
26; 185,1; 186,1.14.16.18.20;
187,29; 188,1.7; 189,14.15;
190,1.2.14.15.16.19.20.33;
191,3.5.17.18.31.32.35;
192,1.3.8.12.25.34; 193,16;
194,23.25.26; 195,2.5.8.12.14.16.20;
196,13; 197,6.9; 198,7; 200,34;
202,3.7.29.30.32; 203,1.26;
204,1.2.9.28.31.32.34;
205,1.9.18.19;
206,5.7.9.10.11.30.33;
207,1.2.5.16.33; 208,8.19.20.25.35;
209,4.15; 210,2; 211,32.33.36;
212,1.2.4.5.7.9.29.31.34; 213,4.5.6;
214,12.13.15.16.18.25.28.30.31.36;
215,1.2; 217,19; 218,28.35;
219,25.30;
220,4.8.11.15.17.18.25.27;
221,3.5.12.14.20.23.29.30.31;
222,2.21.23.25.26.32; 226,5.7.11;
227,14
thermoteros, hotter, 188,7; 202,33.34
thermotês, heat, 179,7.15;
180,2.6.10.31; 182,4.22;
183,11.18.20; 184,1.3.26.31.32.34;
185,1.2.10; 187,1.6.28; 188,5.7.11;
189,6.8.9; 190,28.31.32;
191,2.9.11.16.21.24.28;

192,14.19.22.28.32; 194,13.35;
195,2.11.18; 196,4.6.15.20.22;
197,4.25.26.28; 203,27.28.32;
204,5.8.13; 205,14.15.18; 206,28;
207,7.12.13.14; 211,13; 213,3;
221,26.27.29.30;
222,10.11.12.14.15.16.17.18.33.34.
35; 226,4.17
thlasthai, to be impressed, 216,20.23
thlastos, impressible, 216,19.20.24.31
thlibesthai, to be squeezed, 200,6
thôrax, chest, 226,14
thrauesthai, to be shattered, 216,17
thrausis, fragility, 226,8.15
thraustos, that which can be
shattered, 216,15
threptikos, nutritive, 186,15
thrix, hair, 217,33
thumiasis, the process of fuming,
fumes, 217,22.32.33; 218,12
thumiasthai, to fume, 218,4.7.13
thumiatos, fumable,
217,14.18.20.24.29;
218,2.4.5.18.24.26
tis, someone, something, certain,
182,9; 183,29; 187,2 et passim; *dia
ti*, why, 216,26
titanos, lime, 207,30; 221,32
tmêtikos, piercing, 208,10
tmêtos, splitting, 217,1.2
toioutos, such, of such a sort, of such
a character, 183,25; 187,29; 188,27;
189,1.13.16; 194,11; 196,13
tomê, a cutting, 217,1
trephein, to nourish, 182,27;
186,23.25; 187,1.6.14; 194,27
trophê, nutrition, 182,22;
186,16.17.21.24.26; 187,4.14.15;
188,22; 189,28; 193,32;
194,21.25.26; 197,4.5.8.10
tropos, fashion, way, type, manner,
185,23; 193,15; 195,30; 197,9;
201,3; 206,4.6; 220,26
turos, curd, 211,2.5.10.20
turousthai, to curdle, 194,11

xêrainein, to dry out, 180,28; 182,9;
192,6; 202,1.8.13.14;
203,4.5.7.10.15.22.23.24.25.27.28.29.
31.33; 204,1.2.3.4.6.11.15.16.23;

209,15.17.18.19.20.21;
210,1.2.3.4.8.10.11.12.13.16.17.18;
211,11.12.15.19; 212,17.35;
214,13.25
xêros, dry, 180,5.33; 181,1.2.11.15;
182,8; 183,9.14.15; 190,7.11;
196,13; 198,4.11.15.24.26.29.34.35;
199,2.4.8.11.13.19.22.27.28.30;
202,15.16; 204,31.33.34; 205,9.27;
207,33; 209,27; 212,29.32;
213,7.20.34.37; 217,30;
218,28.34.35; 219,2.5.23; 222,5;
226,28.31; 227,11
xêroteros, drier, 184,13; 188,8;
192,11; 196,21.23
xêrotês, dryness, 179,7.16; 180,2.3.7;
182,6; 190,9; 198,9; 202,18.19;
205,16; 209,17; 218,6

xulinos, wooden, 225,11.12
xulôdês, a woody body, 217,32; 218,30
xulon, wood, 192,23.29.33; 193,5;
212,10.14; 213,11; 215,19; 218,30;
220,2.3; 222,1

zein, to boil, seethe, 184,22.26; 215,1;
222,10; 223,2
zên, to be alive, 223,32
zôion, animal, 181,32; 182,14.22;
183,2.6; 185,8.30; 186,15; 188,4;
191,7.8; 197,10.13; 199,20.21;
213,10; 219,17; 222,13; 223,7.14.30;
224,3.7; 225,24; 226,2.3.20;
227,18.19
zôpurein, to kindle, 184,31.33

Subject Index

References are to the pages of this book.